Supplements to the 2nd Edition of

RODD'S CHEMISTRY OF CARBON COMPOUNDS

ELSEVIER SCIENCE PUBLISHERS B.V.
Sara Burgerhartstraat 25
P.O. Box 211, 1000 AE Amsterdam, The Netherlands

Distributors for the United States and Canada:

ELSEVIER SCIENCE PUBLISHING COMPANY INC.
52, Vanderbilt Avenue
New York, NY 10017

Library of Congress Card Number: 64-4605

ISBN 0-444-42792-9

© Elsevier Science Publishers B.V., 1987

All rights reserved. No part of this publication may be reproduced, stored in a retrieval system or transmitted in any form or by any means, electronic, mechanical, photocopying, recording or otherwise, without the prior written permission of the publisher, Elsevier Science Publishers B.V./Science & Technology Division, P.O. Box 330, 1000 AH Amsterdam, The Netherlands.

Special regulations for readers in the USA — This publication has been registered with the Copyright Clearance Center Inc. (CCC), Salem, Massachusetts. Information can be obtained from the CCC about conditions under which photocopies of parts of this publication may be made in the USA. All other copyright questions, including photocopying outside of the USA, should be referred to the publisher.

Printed In The Netherlands

Supplements to the 2nd Edition of

RODD'S CHEMISTRY OF CARBON COMPOUNDS

VOLUME I

ALIPHATIC COMPOUNDS
★

VOLUME II

ALICYCLIC COMPOUNDS
★

VOLUME III

AROMATIC COMPOUNDS
★

VOLUME IV

HETEROCYCLIC COMPOUNDS
★

VOLUME V

MISCELLANEOUS
GENERAL INDEX
★

Science
QD
251
R68
1964
v. 4
pt. H

ea

9/11/87

Supplements to the 2nd Edition (Editor S. Coffey) of

RODD'S CHEMISTRY OF CARBON COMPOUNDS

A modern comprehensive treatise

Edited by
MARTIN F. ANSELL
Ph.D., D.Sc. (London) F.R.S.C. C. Chem.
Reader Emeritus, Department of Chemistry,
Queen Mary College, University of London, Great Britain

Supplement to

VOLUME IV HETEROCYCLIC COMPOUNDS

Part H:
Six-Membered Heterocyclic Compounds with (*a*) a Nitrogen Atom Common to Two or More Fused Rings; (*b*) One Hetero-atom in Each of Two Fused Rings. Six-Membered Ring Compounds with Two Hetero-atoms from Groups VI B, or V B and VI B of the Periodic Table, respectively. Isoquinoline, Lupinane and Quinolizidine Alkaloids

ELSEVIER
Amsterdam — Oxford — New York — Tokyo   1987

CONTRIBUTORS TO THIS VOLUME

KENNETH W. BENTLEY, M.A., D.Sc., D.Phil., F.R.S.E.
Department of Chemistry, Loughborough University,
Loughborough, Leicestershire LE11 3TU

DEREK T. HURST, M.Sc., Ph.D., C.Chem., F.R.S.C.
Department of Chemistry, Kingston Polytechnic,
Kingston-upon-Thames KT1 2EE

DEREK LEAVER, B.Sc., Ph.D., M.R.I.C., C.Chem., F.R.S.E.
Department of Chemistry, The University,
Edinburgh EH9 3JJ

MALCOLM SAINSBURY, D.Sc., Ph.D., C.Chem., F.R.S.C.
Department of Chemistry, The University,
Bath BA2 7AY

RAYMOND E. FAIRBAIRN, B.Sc., Ph.D., F.R.S.C.
Formerly of Research Department,
Dyestuffs Division,
I.C.I. (INDEX)

PREFACE TO SUPPLEMENT IVH

The publication of this volume continues the supplemen-
tation of the second edition of Rodd's Chemistry of Carbon
Compounds thus keeping this major work of reference up-to-
date. This supplement covers chapters 36-41 inclusive which
appeared in Volume IVH of the second edition. Although each
chapter in this book stands on its own, it is intended that
it should be read in conjunction with the parent chapter in
the second edition.

As editor I have again been fortunate in having four
contributors who have each produced clear, critical and very
readable accounts of the developments in the particular area
of chemistry they have surveyed. I am particularly grateful
to Dr Malcolm Sainsbury who has continued his many valuable
contributions to "Rodd" by writing three of the chapters in
this volume which constitutes over half of the supplement.
Again Dr Fairbairn has enhanced the value of the supplement
by producing an excellent index.

At a time when there are many specialist reviews, mono-
graphs and reports available, there is still in my view an
important place for "Rodd" which gives a broad coverage of
organic chemistry. One aspect of the value of this work is
that it allows the expert in one field to quickly find out
what is happening in other fields of chemistry. On the
other hand, a chemist looking for a way into a field of
study will find in "Rodd" a review of the important aspects
of that field together with key references to specialist
works providing more detailed information.

Like other supplements this volume has been produced by
direct reproduction of each contributor's manuscript. I am
very appreciative of all the effort the authors and their
secretaries have put into producing clear camera-ready manuscripts containing very carefully drawn diagrams. I also wish
to thank the staff at Elsevier for all the help they have
given me and for transforming the manuscripts into a well
produced book.

March 1987   Martin Ansell

CONTENTS

VOLUME IV H

Heterocyclic Compounds: Six-Membered Heterocyclic Compounds with (*a*) a Nitrogen Atom Common to Two or More Fused Rings; (*b*) One Hetero-atom in Each of Two Fused Rings. Six-Membered Ring Compounds, with Two Hetero-atoms from Groups VI B, or V B and VI B of the Periodic Table, respectively. Isoquinoline, Lupinane and Quinolizidine alkaloids

Chapter 36. Alkaloids of the Morphine-Hasubanonine Group
by K.W. BENTLEY

Chapter 37. Fused Heterocyclic Systems having a Nitrogen Atom Common to Two or More Rings
by D. LEAVER

Chapter 41. Compounds Containing a Six-Membered Ring with Two Hetero-
atoms from Groups V and VI, respectively of the Periodic Table. Oxazines,
Thiazines and Their Analogues
by M. SAINSBURY

OFFICIAL PUBLICATIONS

B.P.	British (United Kingdom) Patent
F.P.	French Patent
G.P.	German Patent
Sw.P.	Swiss Patent
U.S.P.	United States Patent
U.S.S.R.P.	Russian Patent
B.I.O.S.	British Intelligence Objectives Sub-Committee Reports
F.I.A.T.	Field Information Agency, Technical Reports of U.S. Group Control Council for Germany
B.S.	British Standards Specification
A.S.T.M.	American Society for Testing and Materials
A.P.I.	American Petroleum Institute Projects
C.I.	Colour Index Number of Dyestuffs and Pigments

SCIENTIFIC JOURNALS AND PERIODICALS

With few obvious and self-explanatory modifications the abbreviations used in references to journals and periodicals comprising the extensive literature on organic chemistry, are those used in the World List of Scientific Periodicals.

LIST OF COMMON ABBREVIATIONS AND
SYMBOLS USED

A	acid
Å	Ångström units
Ac	acetyl
α	axial; antarafacial
as, $asymm.$	asymmetrical
at	atmosphere
B	base
Bu	butyl
b.p.	boiling point
C, mC and μC	curie, millicurie and microcurie
c, C	concentration
C.D.	circular dichroism
conc.	concentrated
crit.	critical
D	Debye unit, 1×10^{-18} e.s.u.
D	dissociation energy
D	dextro-rotatory; dextro configuration
DL	optically inactive (externally compensated)
d	density
dec. or decomp.	with decomposition
deriv.	derivative
E	energy; extinction; electromeric effect; Entgegen (opposite) configuration
E1, E2	uni- and bi-molecular elimination mechanisms
E1cB	unimolecular elimination in conjugate base
e.s.r.	electron spin resonance
Et	ethyl
e	nuclear charge; equatorial
f	oscillator strength
f.p.	freezing point
G	free energy
g.l.c.	gas liquid chromatography
g	spectroscopic splitting factor, 2.0023
H	applied magnetic field; heat content
h	Planck's constant
Hz	hertz
I	spin quantum number; intensity; inductive effect
i.r.	infrared
J	coupling constant in n.m.r. spectra; joule
K	dissociation constant
kJ	kilojoule

LIST OF COMMON ABBREVIATIONS

k	Boltzmann constant; velocity constant
kcal	kilocalories
L	laevorotatory; laevo configuration
M	molecular weight; molar; mesomeric effect
Me	methyl
m	mass; mole; molecule; *meta-*
ml	millilitre
m.p.	melting point
Ms	mesyl (methanesulphonyl)
$[\text{M}]$	molecular rotation
N	Avogadro number; normal
nm	nanometre (10^{-9} metre)
n.m.r.	nuclear magnetic resonance
n	normal; refractive index; principal quantum number
o	*ortho-*
o.r.d.	optical rotatory dispersion
P	polarisation, probability; orbital state
Pr	propyl
Ph	phenyl
p	*para-*; orbital
p.m.r.	proton magnetic resonance
R	clockwise configuration
S	counterclockwise config.; entropy; net spin of incompleted electronic shells; orbital state
S_N1, S_N2	uni- and bi-molecular nucleophilic substitution mechanisms
S_Ni	internal nucleophilic substitution mechanisms
s	symmetrical; orbital; suprafacial
sec	secondary
soln.	solution
symm.	symmetrical
T	absolute temperature
Tosyl	*p*-toluenesulphonyl
Trityl	triphenylmethyl
t	time
temp.	temperature (in degrees centigrade)
tert.	tertiary
U	potential energy
u.v.	ultraviolet
v	velocity
Z	zusammen (together) configuration

LIST OF COMMON ABBREVIATIONS

α	optical rotation (in water unless otherwise stated)
$[\alpha]$	specific optical rotation
α_A	atomic susceptibility
α_E	electronic susceptibility
ε	dielectric constant; extinction coefficient
μ	microns (10^{-4} cm); dipole moment; magnetic moment
μ_B	Bohr magneton
μ_g	microgram (10^{-6} g)
λ	wavelength
ν	frequency; wave number
χ, χ_d, χ_μ	magnetic, diamagnetic and paramagnetic susceptibilities
$\sim$	about
(+)	dextrorotatory
(-)	laevorotatory
(±)	racemic
$\ominus$	negative charge
$\oplus$	positive charge

Chapter 36

ALKALOIDS OF THE MORPHINE-HASUBANONINE GROUP

K.W. BENTLEY

Developments in the chemistry of the alkaloids of this group have been reviewed annually in Specialist Periodical Reports of The Royal Society of Chemistry "The Alkaloids" Vols. 4 (1972-3) to 13 (1981-2) and thereafter in Natural Product Reports 1984, 1, 355, 1985, 2, 81 and 1986, 3, 153.

1. New Morphine Alkaloids

The two diastereoisomeric *N*-oxides of morphine of codeine and of thebaine have been isolated from *Papaver somniferum* (J.D. Phillipson, S.S. Handa and S. El Dabbas (J.Pharm.Pharmacol., 1976, 28 Suppl., 70P), nororipavine (oripavidine) (1) has been found in *Papaver orientale* and identified by *N*-methylation to oripavine and *O,N*-dimethylation to thebaine (I.A. Israilov, Khim.Prir.Soedin., 1977, 714) and 14-hydroxycodeine (2) and 14-hydroxycodeinone have been obtained from *Papaver bracteatum* (H.G. Theuns *et al.*, Phytochem. 1977, 16, 753).
    The unusual carbinolamine ether (6), which has not been given a simple name has been found as a natural product in *Papaver bracteatum*, where it occurs together with the two thebaine *N*-oxides. The major *N*-oxide prepared by oxidising thebaine with hydrogen peroxide has been found to rearrange to (6) on boiling in chloroform, though the principal decomposition product is the phenolic secondary amine (9). The minor *N*-oxide of thebaine is stable under the same conditions. Assuming that the major *N*-oxide has the oxygen in the equatorial position in the nitrogen-containing ring the rearrangement can be depicted as a Cope degradation of (3) to give the hydroxylamine (4) followed by loss of water to give the imine (5) in which a 1,2-hydride shift and nucleophilic attack by the nitrogen at C-5 leads to (6). The formation of the major product (9) can be explained by postulating the rearrangement of (4)

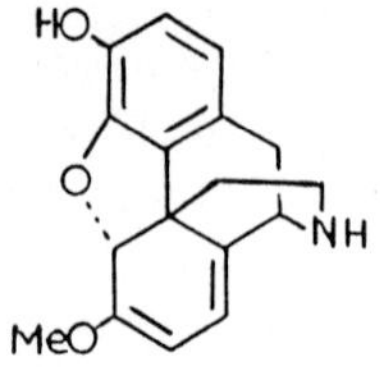

(1) 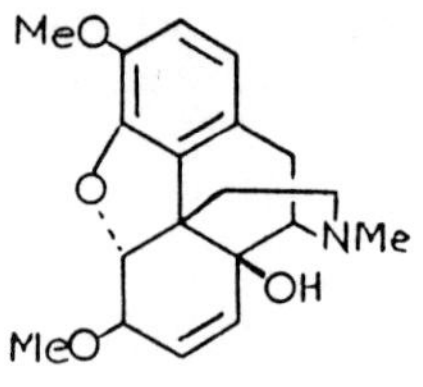   (2) 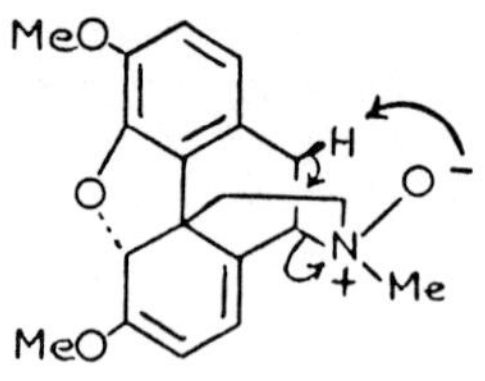   (3)

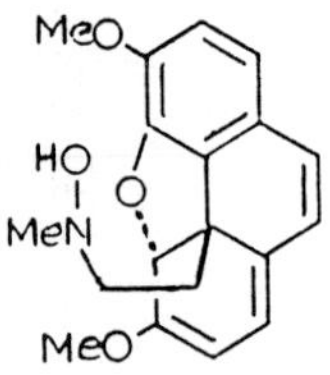

(4) 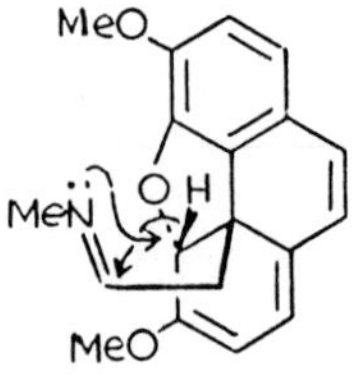   (5) 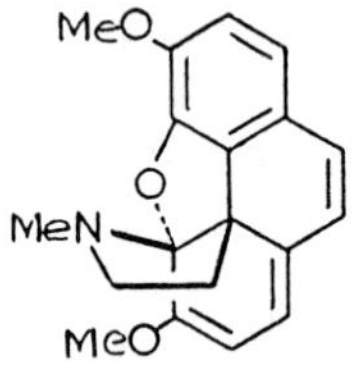   (6)

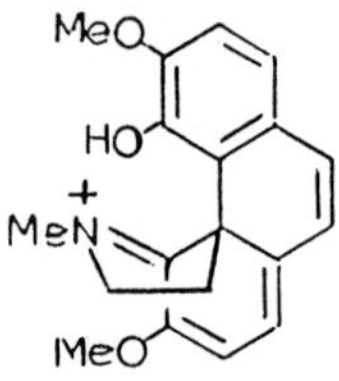

(7) 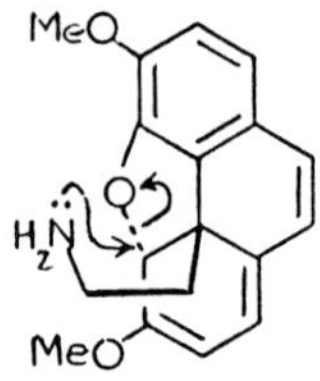   (8) 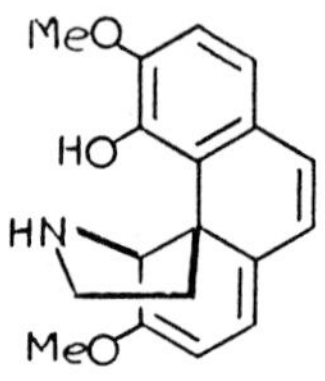   (9)

with loss of formaldehyde to (8), followed by nucleophilic
opening of the 4,5-oxide bridge (H.G. Theuns *et al.*, J.chem.
Soc., Perkin I, 1984, 1701). Spectroscopic studies support
the assignment of the configuration (3) to the major *N*-oxide
of thebaine (H.G. Theuns *et al.*, Org.mag.Res., 1984, 22, 793).
In acid solution (6) is converted into the iminium salt (7)
which yields unchanged (6) on basifaction; the enol ether
system is resistant to hydrolysis.
  Neodihydrothebaine (10, R^1=OMe, R^2=H) and the isomeric bra-

ctazonine (10, R¹=H, R²=OMe) have also been isolated from *Papaver bracteatum*. The former has been shown to be identical
with the product previously obtained by the treatment of thebaine with anhydrous magnesium iodide (which gives the iodomag-
nesium salt of the imine (15, R¹=OMe, R²=H) followed by reduction with lithium aluminium hydride and both alkaloids have

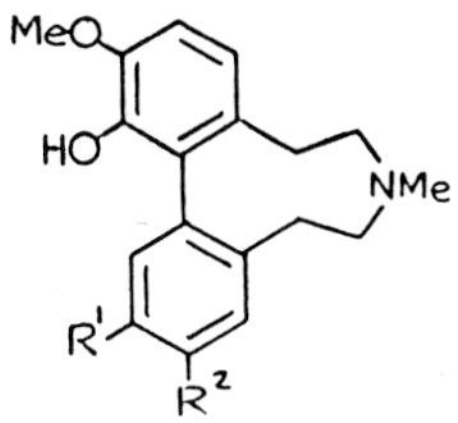

(10)

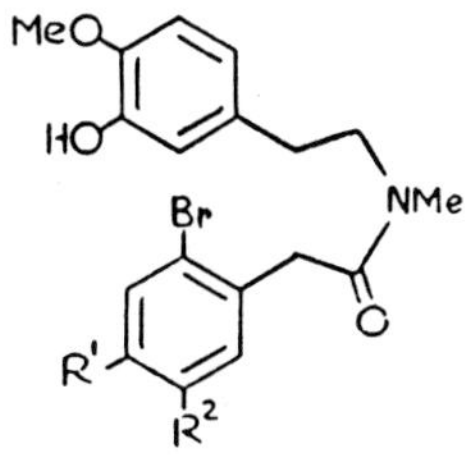

(11)

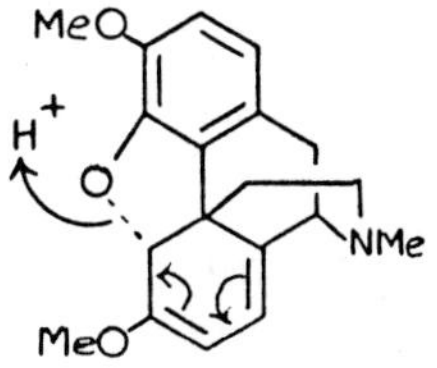

(12)

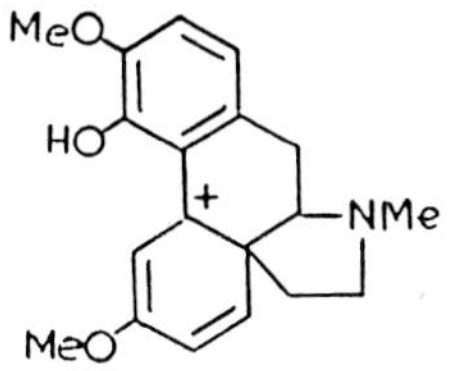

(13)

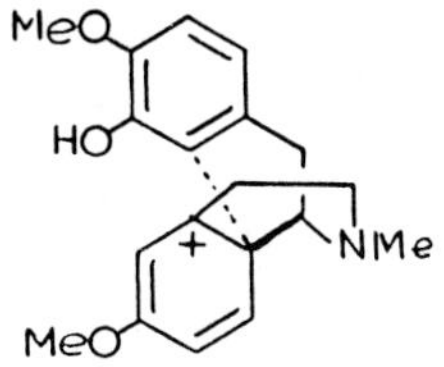

(14)

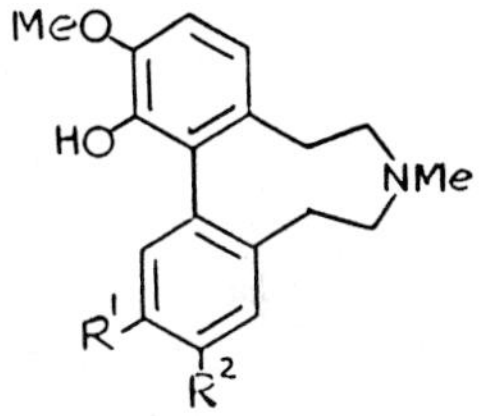

(15)

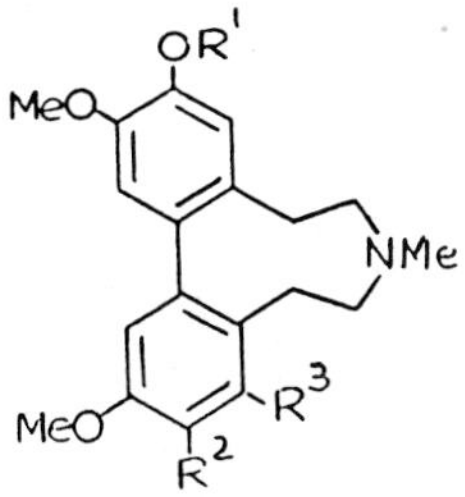

(16)

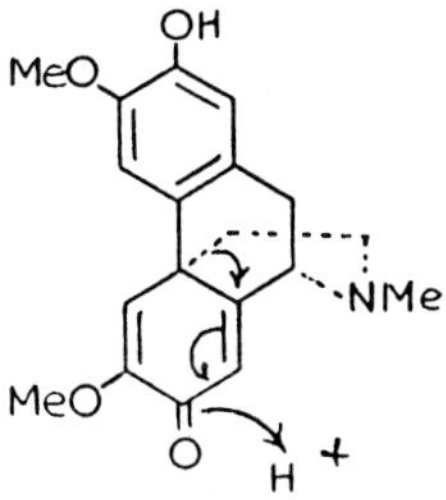

(17)

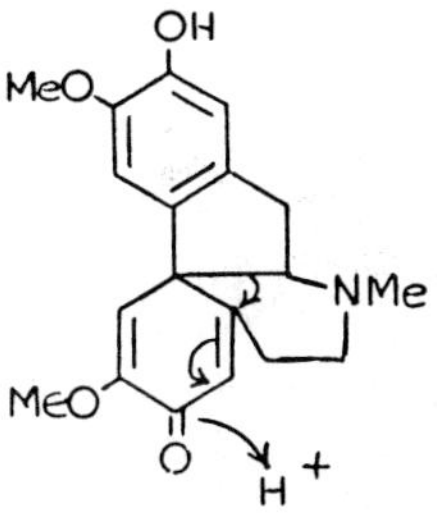

(18)

been synthesised by the photocyclisation of the amides (11, R^1 =OMe, R^2=H) and (11, R^1=H, R^2=OMe) (H.G. Theuns *et al.*, Phytochem., 1984, 23, 1157; Heterocycles, 1984, 22, 2007). Although other biogenetic pathways are possible, it is almost certain that these alkaloids arise from alternative rearrangements of thebaine (12) by migration of either the ethanamine chain or the aromatic nucleus to give (13) and (14), which then collapse to the imines (15, R^1=OMe, R^2=H) and (15, R^1=H, R^2=OMe), followed by reduction. It may be noted that analogues of these bases occur naturally as the alkaloids laurifinine (16, R^1=R^2=R^3=H), laurifonine (16, R^1=Me, R^2=R^3=H), erybidine (16, R^1=R^3=H, R^2=OMe) and protostephanine (16, R^1=Me, R^2=H, R^3=OMe). Of these erybidine could arise by a mechanistically similar rearrangement of pallidine (17) and laurifinine and laurifonine could be formed from the related secondary alcohol. An alternative biogenesis is possible from the dienone (18) through the same intermediate, but it may be noted that (17) is derived by the oxidative coupling of the known alkaloid reticuline, whereas (18) would arise from the unknown protosinomenine.

2. *O- and N-Demethylations*

The *O*-methylation of morphine to codeine has for long been practised in very high yield on a manufacturing scale, but the reverse demethylation of codeine to morphine is difficult because of the instability of both alkaloids under acid conditions. This demethylation has been claimed in 90% yield using boron tribromide in chloroform (K.C. Rice, J.med.Chem., 1977, 20, 164) and in 80% yield using potassium t-butoxide in propanthiol (J.A. Lawson and J.I. De Graw, J.med.Chem., 1977, 20, 165). In view of the pharmacological and clinical importance of derivatives of normorphine new methods of *N*-demethylation are constantly being sought. Improvements on the traditional methods using cyanogen bromide, ethyl chloroformate and ethyl azodicarboxylate followed by hydrolysis have been claimed using methyl chloroformate followed by hydrazine (G.A. Brine *et al.*, Org.Prep.Proc.Internat., 1976, 8, 103), vinyl chloroformate (R. Olafson *et al.*, Tetrahedron Letters, 1977, 1567, 1571) and α-chloroethylchloroformate (R. Olofson *et al.*, J.org.Chem. 1984, 49, 2081). Since esters and urethanes containing the vinyloxycarbonyl group do not hydrolyse at the same rate, the urethane (19, R=CO.OCH=CH$_2$), obtained from morphine and an excess of vinyl chloroformate, can be hydrolysed to (19, R=H) and treatment of this with allyl bromide, followed by hydrolysis in aqueous acid, yields *N*-allylnormorphine in 84% overall yield from morphine. The reaction products of morphine and codeine with α-chloroethyl chloroformate can be converted into the secondary bases simply by heating in methanol.

(19)   (20)   (21)

(22)   (23)   (24)

3.   *Codeine and its Reactions*

Following the discovery and commercial cultivation of strains of *Papaver bracteatum* in which alkaloid biosynthesis ceases at thebaine, with the accumulation of large amounts of this base and no morphine in the plant, the conversion of thebaine into codeine has become of great commercial importance. The key step is the production of codeinone, which is reduced to codeine in high yield. When thebaine is treated with anhydrous hydrogen bromide in methylene chloride, followed by methanol the product is 8β-bromodihydrocodeinone dimethylketal (20), which can be hydrolysed in acetone to codeinone (24) (J.P. Gavard *et al*., Bull.Soc.chim.Fr., 1965, 486) and an improved preparation giving yields >90% omits the methanol treatment and effects dehydrobrominisation of (23, R=Br) with aqueous sodium bicarbonate (Fabrica de Productos Quimicos y Farmaceuticos Abello S.A. Belg.Pat. 839,732; Chem.Abs. 1971, <u>87</u>, 6241). An alternative process claimed to give 85% of codeine from thebaine without the purification of intermediates involves the treatment of thebaine with mercury (II) acetate in methanol to give the ketal (21, R=HgOAc) which can be converted into neopinone (22) by 3M acetic acid or by reduction with sodium borohydride to (21, R=H) followed by hydrolysis. Treatment of neopinone with anhydrous hydrogen chloride or bromide

(25)

(26)

(27)

(28)

(29)

(30)

gives (23, R=Cl) or (23, R=Br), which can be converted into codeinone (24) as before (R. Barber and H. Rapoport, J.med. Chem., 1976, 19, 1175).

Codeinone is oxidised by alkaline hydrogen peroxide to the 7,8-β-epoxide (25), which can be reduced by sodium borohydride to codeine epoxide (M.P. Kotick, J.med.Chem., 1981, 24, 722). Treatment of codeinone with diazomethane affords 7,8-β-methano-dihydrocodeinone (26), which is cleaved by hydrochloric acid to 8β-chloromethyldihydrocodeinone (27, R=Cl), reducible to the 8β-methyl compound (27, R=H) (Kotick, loc cit.). 8β-Alkyl-dihydrocodeinones (27, R=H, Me, Et) are obtained by treating codeinone with lithium dialkyl cuprates; in some cases small amounts of the 8α-alkyl isomers are also produced. 8β-Acyl-dihydrocodeinones can be formed via their enol ethers by using the equivalent of an acyl anion, for example with lithium di-(α-ethoxyvinyl)cuprate codeinone gives the enol ether (28), which can be hydrolysed to 8β-acetyldihydrocodeinone. (Kotick et al., J.med.Chem., 1980, 23, 166). In derivatives of 14-hydroxycodeinone increased steric hindrance at C-8 results in the formation of greater quantities of 8α-alkyl compounds for example (29, R=Me), (29, R=SiMe$_3$) and (29, R=SiMe$_2$Bu^t) give 66%, 80% and 100% of the 8α-methyldihydrocodeinone analogue on treatment with lithium dimethylcuprate (D.L. Leland, J.O. Polazzi and Kotick, J.org.Chem., 1981, 46, 4012).

Codeinone undergoes Michael addition reactions with anions derived from nitromethane and diethyl malonate to give (30, R^1 =H, R^2=NO$_2$) and (30, R^1=R^2=COOEt) respectively (Kotick and Polazzi, J.heterocyclic Chem., 1981, 18, 1029).

4. $Dihydrocodeinone$ and its $Reactions$

Dihydrocodeinone reacts with formaldehyde to give 7,7-di(hyd-roxymethyl)dihydrocodeinone (31), which has been converted in-to the corresponding 7,7-disubstituted derivatives of codeine (32) and isocodeine (33). The bis toluenesulphonyl esters of both (32) and (33) have been subjected to the same sequence of transformations. Hydrolysis of the diester of (33) with sod-ium hydroxide in butanone gives (34, R=CH$_2$OTOS), which is hyd-rolysed and oxidised to the aldehyde (34, R=CHO). This is converted by the Wittig reaction into the olefin (35), which is reduced to the tetrahydrocompound. The dihydrocodeine an-alogue (32) gives similar compounds with the four-membered ether ring on the α-face of the molecule and demethylation of the analogue of tetrahydro-(35) with hydrobromic acid involves ether fission to give (36, R=Br) reducible to (36, R=H). The

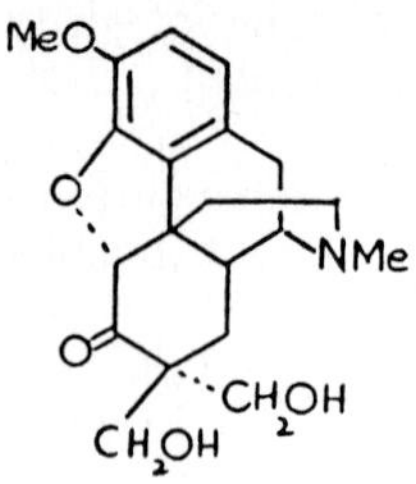

(31)

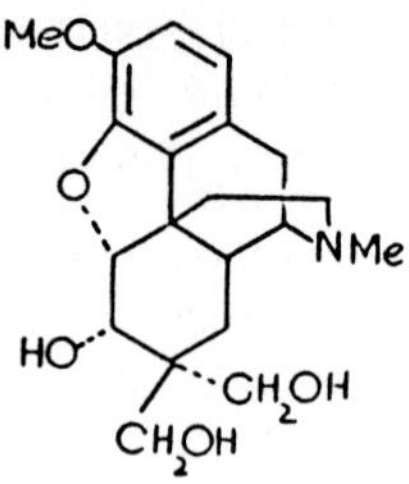

(32)

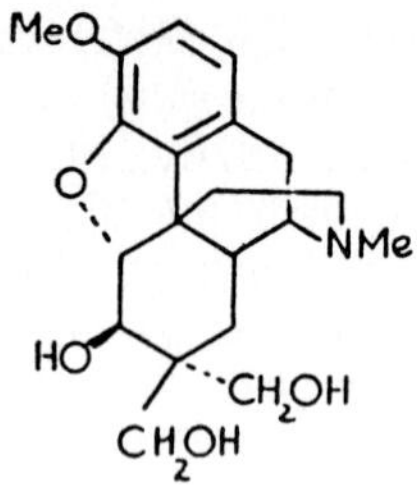

(33)

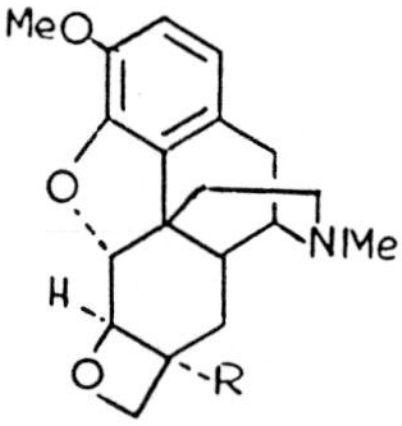

(34)

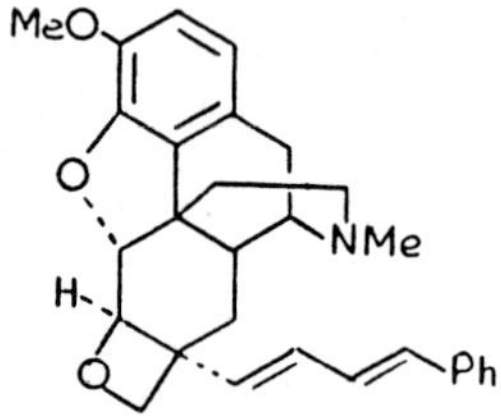

(35)

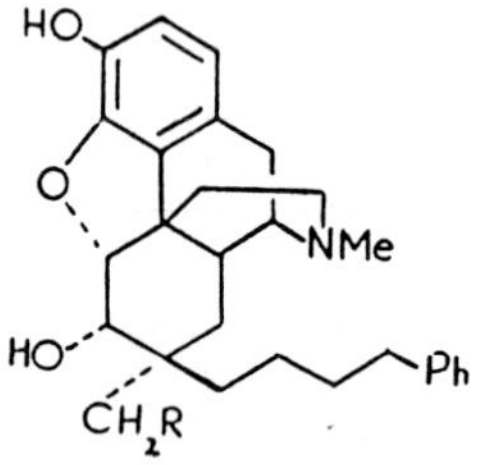

(36)

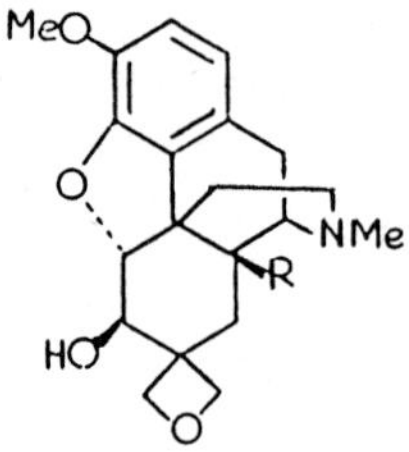

(37)

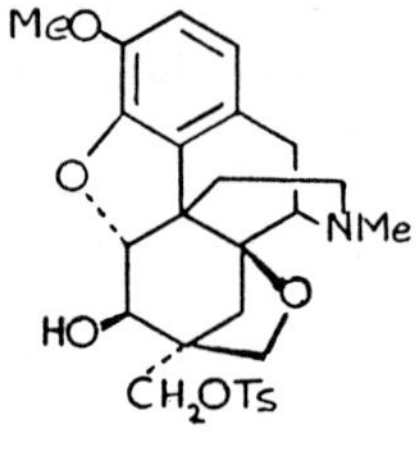

(38)

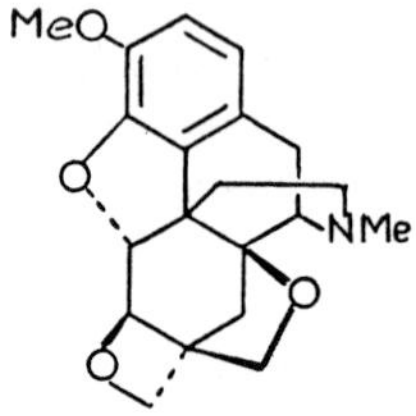

(39)

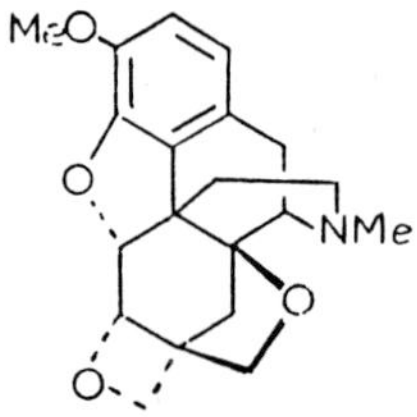

(40)

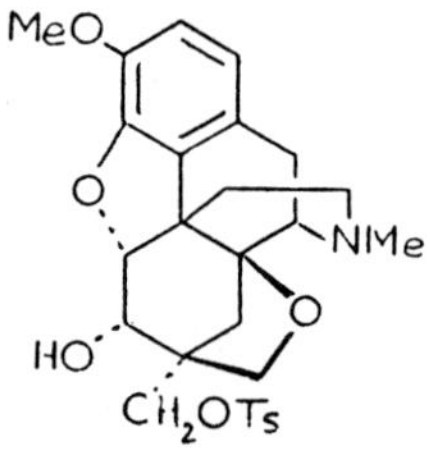

(41)

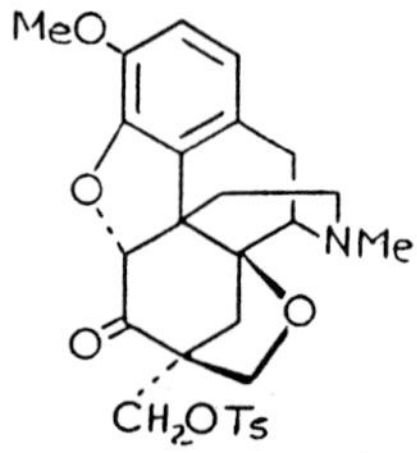

(42)

triol (33) gives two mono-toluenesulphonates, one of which on
treatment with base gives (34, R=CH$_2$OH) whereas the other gives the cyclic ether (37, R=H). (Leland and Kotick, J.org.Chem.,
1983, 48, 1813).

14-Hydroxycodeinone reacts with formaldehyde to give the
14-hydroxy analogue of (31) which has been converted into 14-
hydroxy-(33). Since five-membered cyclic ethers are more easily formed than four-membered ethers this compound behaves
differently from (33) under similar conditions. With toluenesulphonyl chloride it yields a monotosylate (from which (37,
R=OH) can be prepared) and a ditosylate, which is converted
into (38) by ammonia. Sodium hydroxide converts (38) into
an ether assigned the highly-strained structure (39), since it
is not identical with (40), prepared by the action of base on
(41), itself prepared by reduction of the ketone (42) obtained
by the oxidation of the alcohol (38) (Kotick, J.org.Chem.,
1983, 48, 1819).

Photolysis of dihydrocodeinone in aqueous hydrochloric acid
yields 5α-hydroxydihydrothebainone (43, R=H); in methanol,
ethanol and acetic acids the products are (43, R=Me), (43, R=
Et) and (43, R=Ac) respectively but in diethylamine 5β-diethyl-
aminodihydrocodeinone (44) and 5β-diethylaminodihydrothebain-
one are formed (M. Boess and W. Fleischhacker, Ann. 1981,
2002).

Elimination of methanol from dihydrocodeinonedimethyl ketal
yields dihydrothebaine (46) which is converted by *N*-bromoacet-
amide in methanol into 7-bromodihydrocodeinone dimethyl ketal
(47). Potassium *t*-butoxide in dimethyl sulphoxide at 120°
converts this into thebaine, but at 60°C the product is codeinone dimethyl ketal (48), which can be hydrolysed to codeinone.

(43)   (44)   (45)

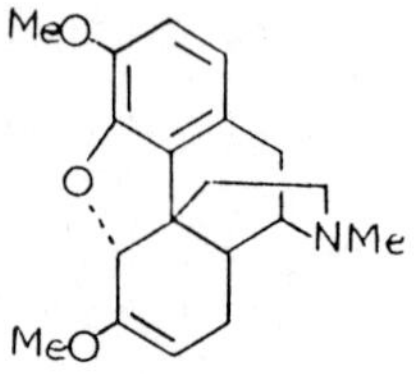

(46) 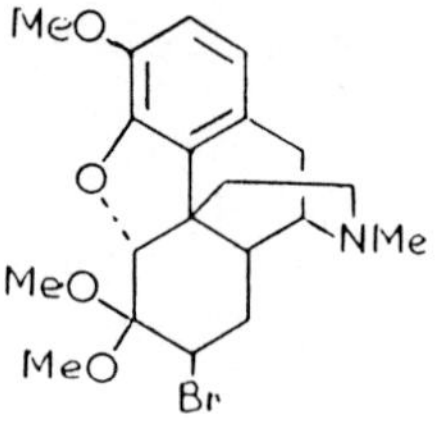   (47) 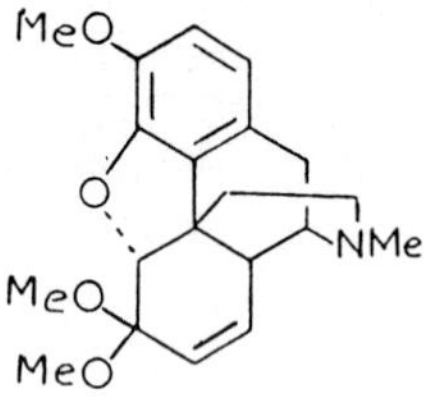   (48)

Since dihydrocodeinone can be prepared easily from dihydro-thebainone, which can be synthesised in reasonable yield, these processes may have commercial potential (D.D. Weller and Rapoport, J.med.Chem., 1975, 19, 1171).

Treatment of dihydrocodeinone with dimethylsulphoxonium methylide affords the epoxide (49) from which the substituted dihydrocodeines (50, R=H), (50, R=Cl), (50, R=OH) and (50, R= N₃) have been prepared. The isomeric dihydroisocodeines can be parepared from the isomeric epoxide, which is accessible from the olefin (51) (G. Horvath et al., Magy.Kem.Foly., 1976, 82, 418).

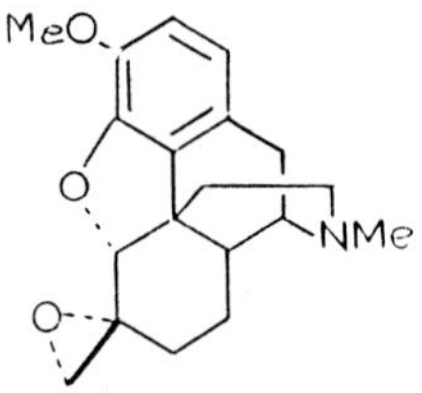

(49) 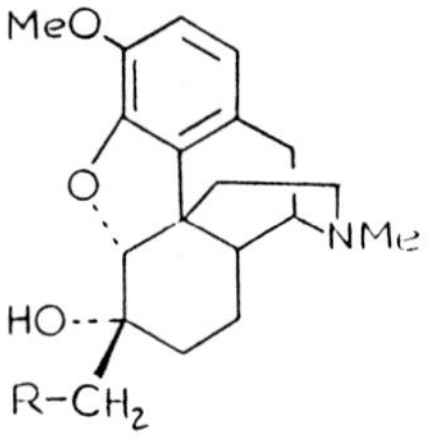   (50) 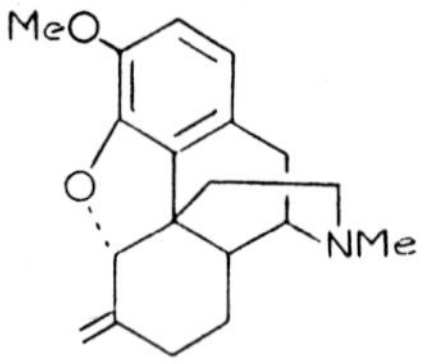   (51)

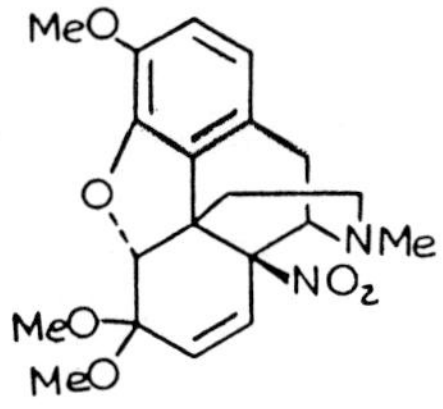

(52)

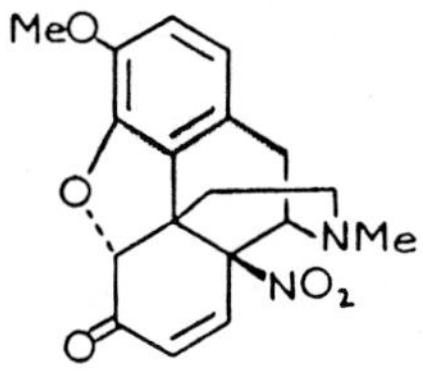

(53)

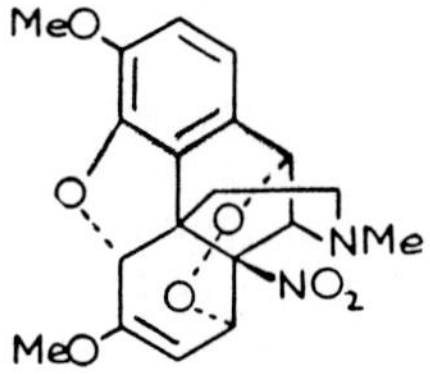

(54)

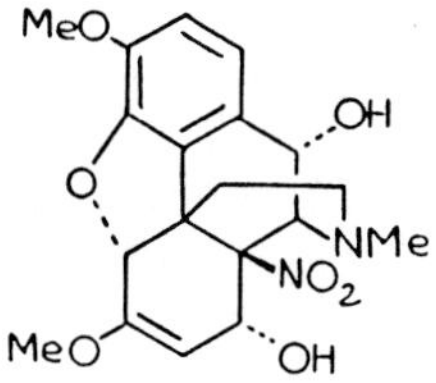

(55)

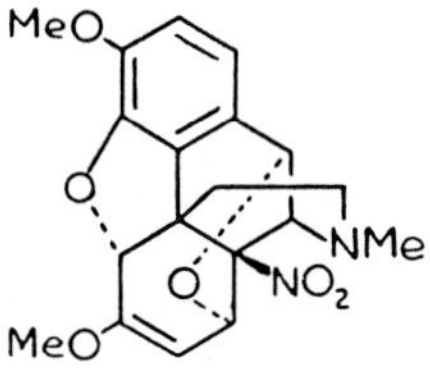

(56)

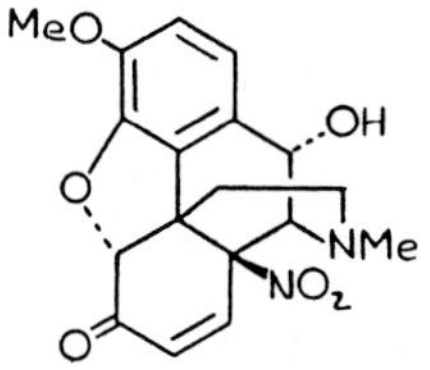

(57)

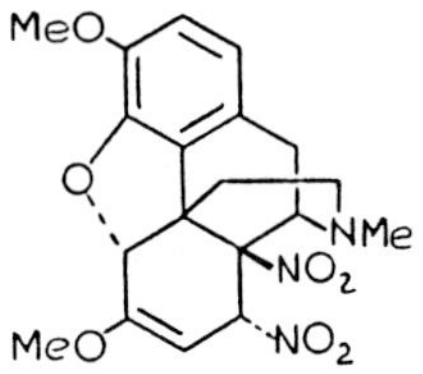

(58)

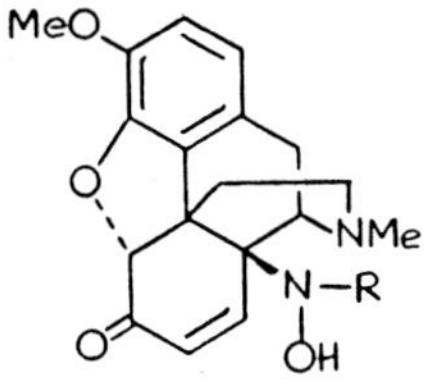

(59)

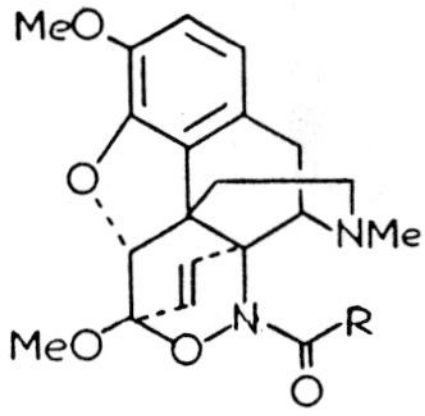

(60)

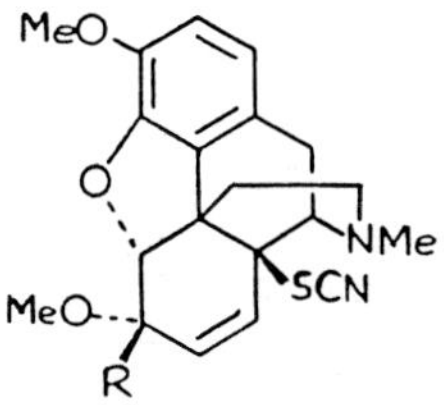

(61)

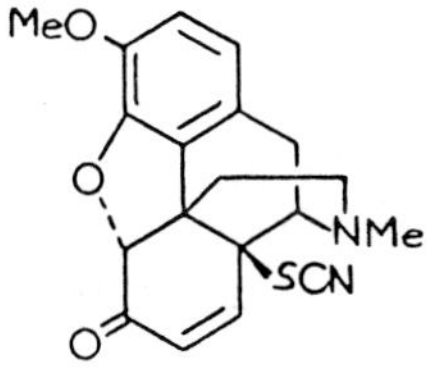

(62)

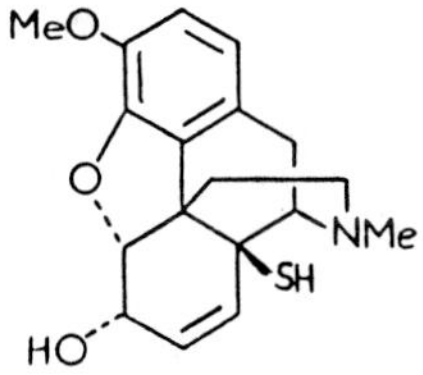

(63)

5. 14-Substituted Codeinones

Treatment of thebaine with tetranitromethane in methanol affords 14-nitrocodeinone dimethyl ketal (52), which can be reduced and hydrolysed to 14-aminocodeinone (53). Some *N*-alkyl
and *N*-acyl derivatives of this amine have very high analgesic
potency. The alkaloid reacts with tetranitromethane in the
presence of dry oxygen, however to give the peroxide (54),
which is reduced by sodium iodide in acetic acid to the diol
(55) and by triphenylphosphine to the cyclic ether (56), which
is hydrolysed by acid to 10-hydroxy-14-nitrocodeinone (57)
(R.M. Allen, G.W. Kirby and D.J. McDougall, J.chem.Soc., Perkin I, 1981, 1143). 14-Nitrocodeinone (53) is also obtained,
together with 8-nitrothebaine by the action of dinitrogen
tetroxide on thebaine, presumably by initial addition of the
oxide to give (58), followed by alternative eliminations (S.
Archer and P. Osei-Gyimah, J.heterocyclic Chem., 1979, 16,
389). 14-Hydroxylaminocodeinone (59, R=H) is obtained from
thebaine and 1-chloro-1-nitrosocyclohexane (P. Horsewood and
Kirby, J.chem.Res.(S), 1980, 401) and by hydrolysis of the
Diels-Alder adducts (60, R=Me) and (60, R=Ph) obtained from
thebaine and *N*-acylhydroxylamines in the presence of tetraeth-
ylammonium periodate, which presumably generates the nitroso-
compounds (Kirby and G.W. Sweeny, J.chem.Soc., Perkin I, 1981,
1143). Similarly (59, R=Ph) results from the hydrolysis of
the adduct (60, COR=Ph) of thebaine and nitrosobenzene (Kirby
et al., J.chem.Soc., Perkin I, 1979, 3064).
   Thebaine also reacts with dithiocyanogen $(SCN)_2$ to give
(61, R=SCN), which is converted into the ketal (61, R=OMe) by
methanol and into the codeinone (62) by sodium bicarbonate.
Reduction of (62) by lithium aluminium hydride yields 14-mer-
captocodeine (63) (Osei-Gyimah and Archer, J.med.Chem., 1980,
23, 162).
   When 7-hydroxyneopinone dimethyl ketal (64) is heated with
N,*N*-dimethylacetamide dimethyl ketal in xylene at 160°C it undergoes Claisen-Eschenmoser rearrangement to the amide (65),
which can be reduced to the amine (66, R=NMe$_2$) by lithium aluminium hydride and to the alcohol (66, R=OH) by lithium tri-
ethylaluminium hydride. Treatment of the amine with methylio-
dide and of the alcohol with sodium acetate and acetic anhydride gives the quaternary salt (67) which is converted into
the olefin (68) by Hofmann degradation, into the 14-ethyl-
compound (66, R=H) by Emde reduction and into the thioether
(69) by lithium hydride and propanethiol. Reduction of the
amide with sodium aluminium hydride affords (66, R=NMe$_2$), (66,

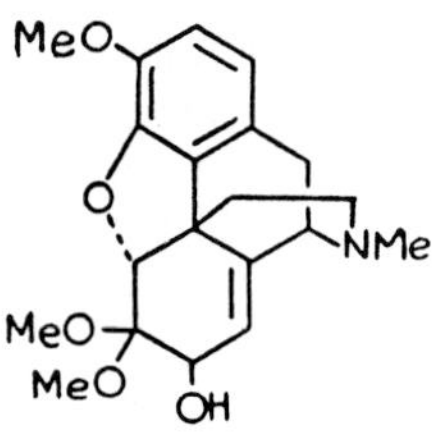

(64) 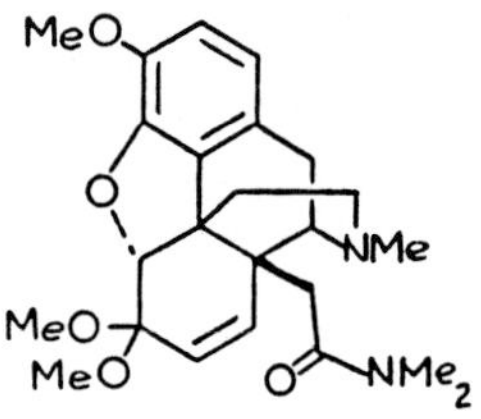   (65) 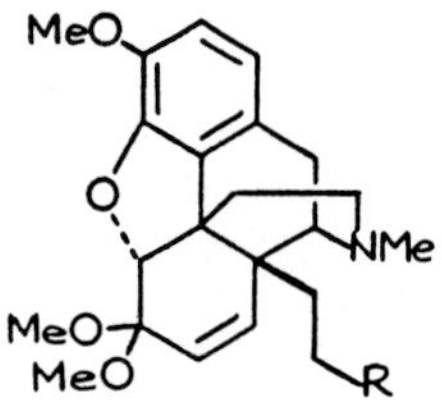   (66)

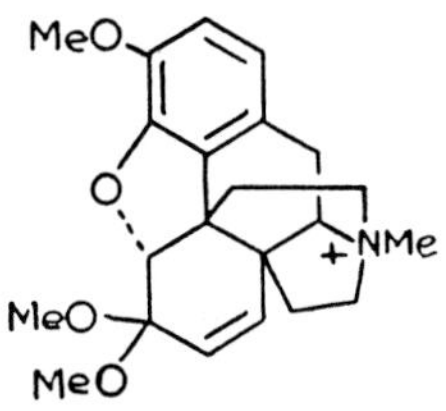

(67) 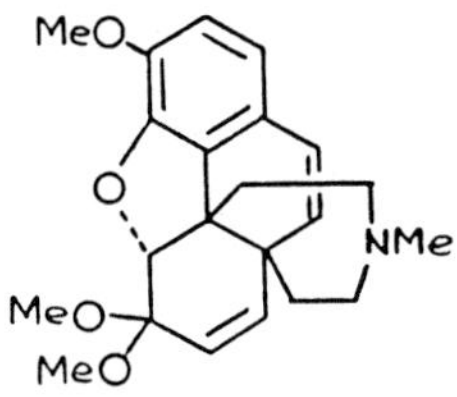   (68) 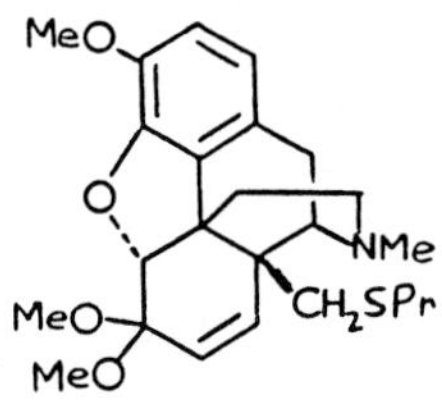   (69)

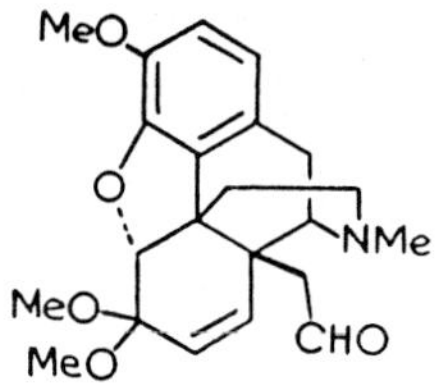

(70) 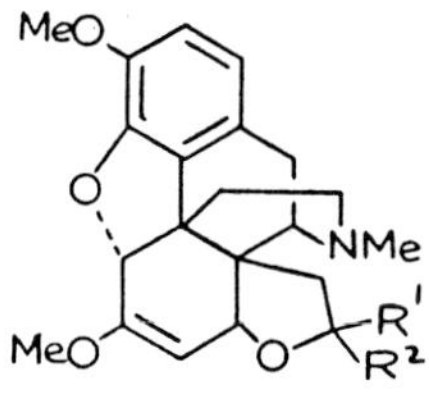   (71) 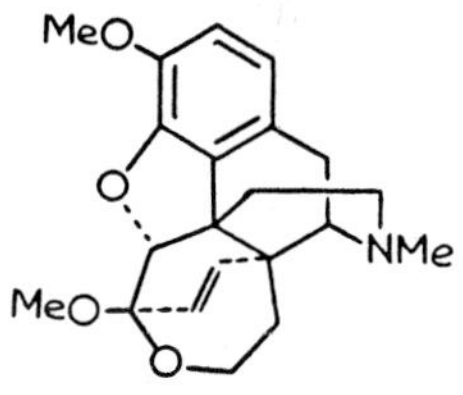   (72)

R=OH) and the aldehyde (70) which is decarbonylated to 14β-methylcodeinone dimethyl ketal. Hydrolysis of most of these ketals with 0.1M hydrochloric acid affords the related derivatives of codeinone, but with more concentrated acid addition to the double bond occurs giving (71, R^1R^2=O) and the related ketone from (65) and (71, R^1=R^2=O), the related ketone and the mixed ketal (72) from the alcohol (66, R=OH) (Fleischhacker and B. Richter, Ber., 1979, 112, 2539, 3054).

The indolinocodeinone (73) is converted by alkalis, presum-

ably *via* the aziridinium salt (74) into 7,14-cyclodihydro-
codeinone (75) which on catalytic reduction yields (76) which
is converted by diazomethane into 14β-methyldihydrocodeinone
(77), the oxiran (78) and further ring expansion products.
The methyl compound (77) is identical with the base prepared
through 14β-methylcodeinone from the aldehyde (70) (Fleischhacker and Klement, Monatsh., 1976, 107, 1029; 1977, 108,
1441).

(73)   (74)   (75)

(76)   (77)   (78)

6. *Thebaine and its Analogues*

The two phenanthrene derivatives (80) and (81), which were
originally   isolated as trace impurities in illicit heroin
samples, have been prepared by heating thebaine with acetic
anhydride; they presumably arise from degradation of thebaine
to (79) followed by the alternative rearrangements shown and

loss of a proton (A.C. Allen *et al.*, J.org.Chem., 1984, 49, 3462). Thebaine has been degraded to thebaol ether (82) simply by boiling it in epichlorohydrin (T.S. Manohasan *et al.*, Indian J.Chem., 1984, 23B, 558). Such degradation had only previously been observed with quaternary salts in acetic anhydride and sodium acetate. When thebaine is treated with butyl lithium at -78°C it gives the C-5 anion which is methylated by methyl fluorosulphonate to 5β-methylthebaine (83) and this has been converted into 5β-methylcodeinone, 5β-methyldihydrocodeinone and Metopon (84) (R.M. Boden *et al.*, J.org.Chem., 1982, 47, 1347).

(79)   (80)   (81)

(82)   (83)   (84)

In contrast with its reaction with Grignard reagents thebaine reacts with lithium dimethylcuprate to give 7β-methyldihydrothebaine-ϕ (85), which can be hydrolysed to 5-methylthebainone-A (86) and its C-14 epimer. Reduction of the phenyl ether of (85) gives the 4-deoxy-analogue, which itself reacts with lithium dimethylcuprate to give (after hydrolysis) the isomeric ketones (87) and (88) (Leland, Polazzi and Kotick, J.org.Chem., 1980, 45, 4026).

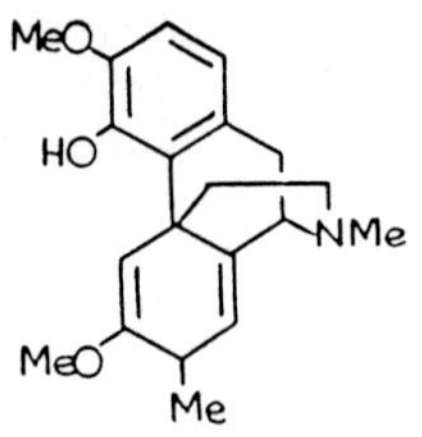

(85)

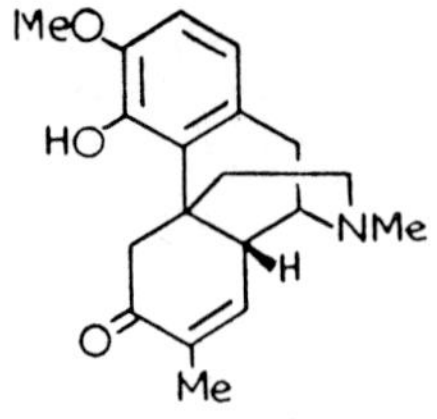

(86)

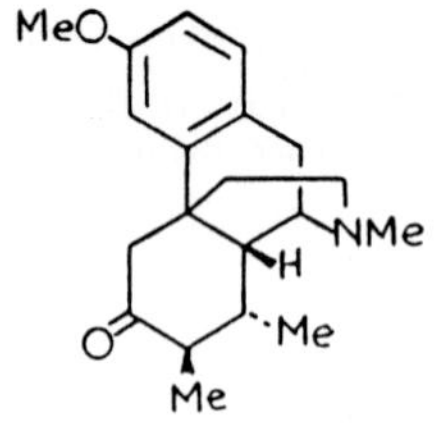

(87)

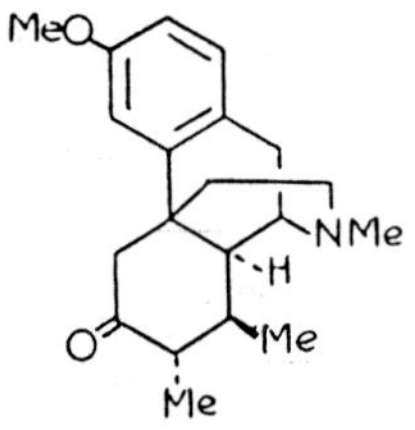

(88)

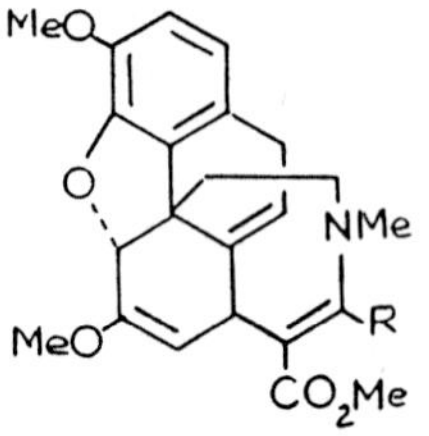

(89)

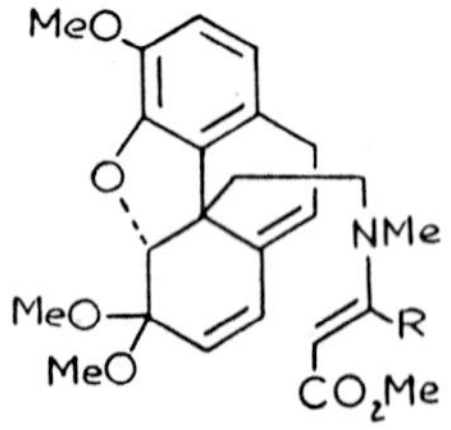

(90)

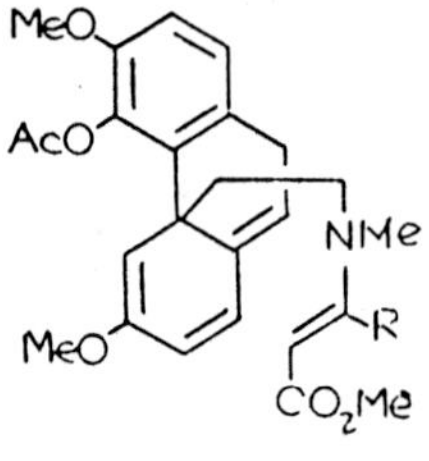

(91)

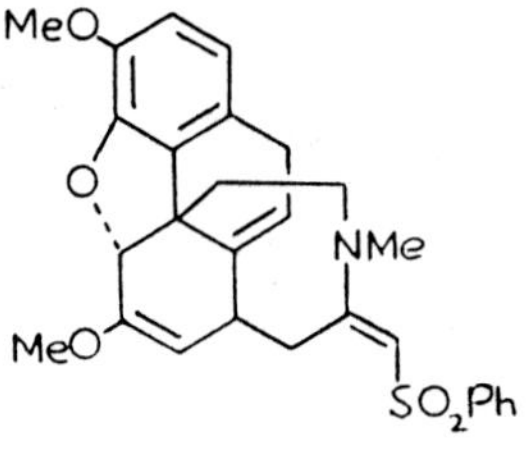

(92)

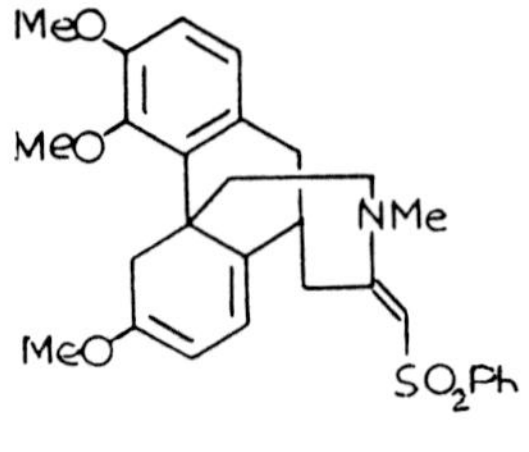

(93)

Thebaine undergoes normal Diels-Alder reactions with acetylenic dienophiles in non-polar solvents but in polar solvents (89, R=H) and (89, R=COOMe) are formed and in methanol the ketals (90, R=H) and (90, R=COOMe) are produced. Under similar conditions *O*-acetyl-β-dihydrothebaine gives the enol ethers (91) and the related dimethyl ketals (K. Hayakawa, I. Fujii and K. Kanematsu, J.org.Chem., 1983, 48, 166; A. Singh *et al.*, J.org.Chem., 1983, 48, 173). Under similar conditions thebaine and *O*-methyl-β-dihydrothebaine react with allenyl phenyl sulphone to give (92) and (93) respectively (Fujii *et al.*, Chem.Comm., 1984, 844).

6-Demethoxythebaine (94, R=H) has been prepared by elimination reactions from methanesulphonyl esters of codeine, isocodeine and neopine and similarly from bromocodide (S. Berenyi *et al.*, Acta Acad.Sci.Hung., 1980, 103, 365; C.W. Hutchins *et al.*, J.med.Chem., 1981, 24, 773; H.C. Beyerman *et al.*, Recl.J.R.Neth.chem.Soc., 1984, 103, 112) and 6-deoxythebaine (94, R=Me) has similarly been prepared from 6-methylcodeine methyl ether (L. Knipmeyer and Rapoport, J.med.Chem., 1985, 28, 461).

6-Demethoxythebaine undergoes Diels-Alder reaction with methyl vinyl ketone to give the 7α-*endo*-adduct (95, R=H) and this ketone has been converted into 6-demethoxyetorphine (96, R=H), which is an analgesic of potency similar to that of etorphine (96, R=OMe) 6-Deoxythebaine with methyl vinyl ketone gives mainly (95, R=Me) together with its 7β-epimer, the *exo* —adduct (97) and the C-8 ketone (98) and of these (95, R=Me) has been converted into (96, R=Me). 6-Demethoxy-β-dihydrothebaine gives the *exo*-adduct (99) (knipmeyer and Rapoport, *loc. cit.*; P.R. Crabbendam *et al.*, Recl.J.R.Neth.chem.Soc., 1984, 103, 296).

(94) (95) (96)

(97)   (98)   (99)

7. Alkaloids of the Sinomenine-Salutaridine Group

The following new alkaloids of this group have been isolated:
Episinomenine (100), which can be epimerised to sinomenine,
from *Ocotea brachybotra* (V.Vecchietti, C. Cassagrande
and G. Ferrari, Tetrahedron Letters, 1976, 1633)
Carococculine (101) from *Cocculus carolinus* (Slatkin *et al.*,
Lloydia, 1974, 37, 488.
Stephodeline (102) and its C-14 epimer, isostephodeline,
from *Stephania delovayi* (M.E. Perel'son, I.I. Fadeeva
and T.N. Ilinskaya, Khim.Prir.Soedin., 1975, 188.
Tridictophylline (103), from *Triclisia dictyophylla* (A.L.
Spiff *et al.*, J.nat.Prod., 1981, 44, 160
Monocrispatine (104), from *Monodera crispata* (A.L. Djakawe,
Ann.Univ.Abidjan, Ser.C, 1981, 17, 105
O-Methylfalvinantine (105) from *Papaver bracteatum* (H.
Meshulam and D. Lavie, Phytochem., 1980, 19, 2633) and
Rhiziocarpa racimifera (D. Dwuma-Badn *et al.*, J.nat.
Prod., 1980, 43, 123)
Sebiferine (106, R=Me), the enantiomer of *O*-methylflavinan-
tine, from *Ocotea acutanglea* (Vecchietti *et al.*, J.chem.
Soc., Perkin I, 1981, 578.
Norpallidine (106, R=H) from *Fumaria vaillantii* (M. Shamma
et al., Phytochem. 1976, 18, 1802)
Pallidinine (197, R=H) and *O*-methylpallidinine (107, R=Me)
from *Ocotea acutanglea* (Vecchietti *et al.*, *loc.cit.*)
Amurinine (108), its C-14 epimer epiamurinine and dihydro-
nudaurine (109), from *Papaver spicatum* (R. Hocquemiller
et al, J.nat.Prod., 1984, 47, 342
Ocobotrine (110), which is the C-14 epimer of dihydrosino-

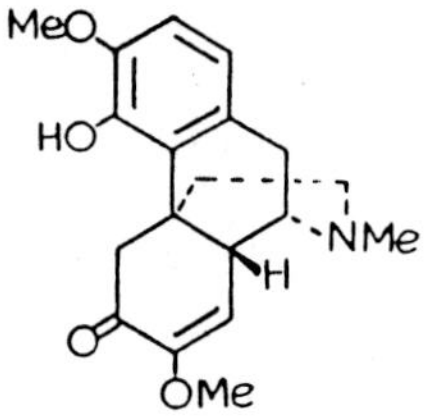

(100)

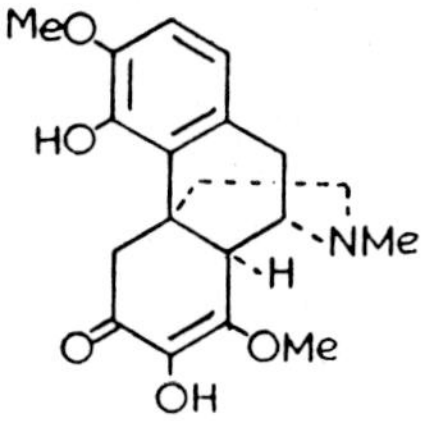

(101)

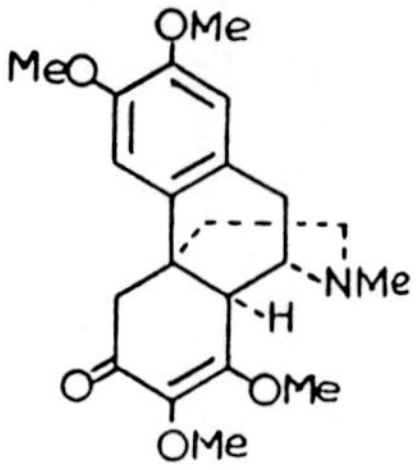

(102)

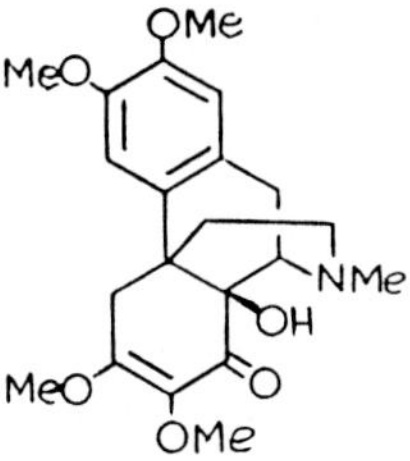

(103)

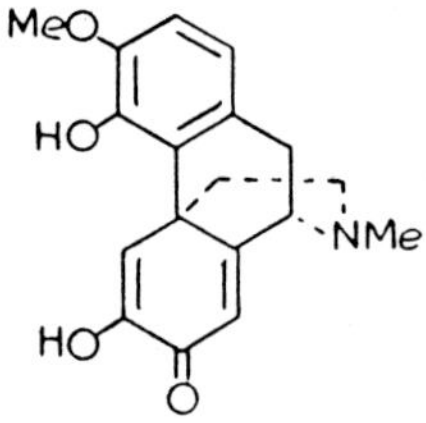

(104)

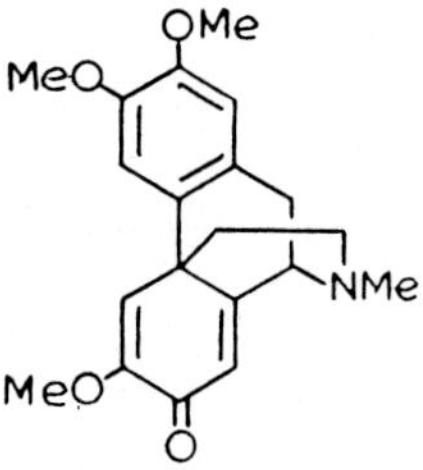

(105)

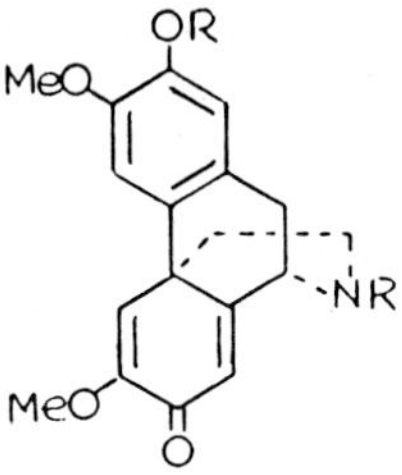

(106)

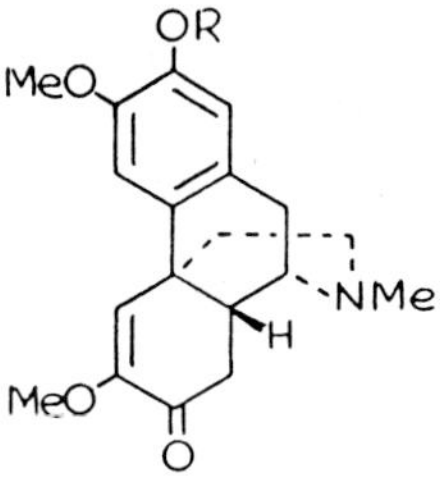

(107)

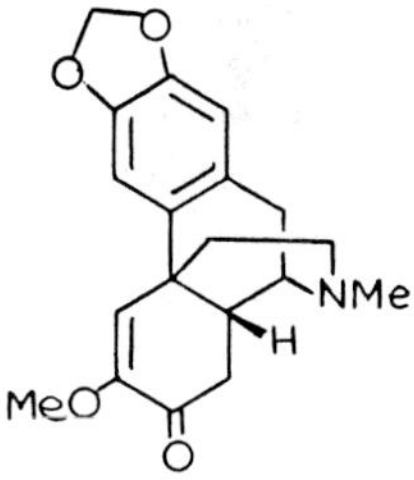

(108)

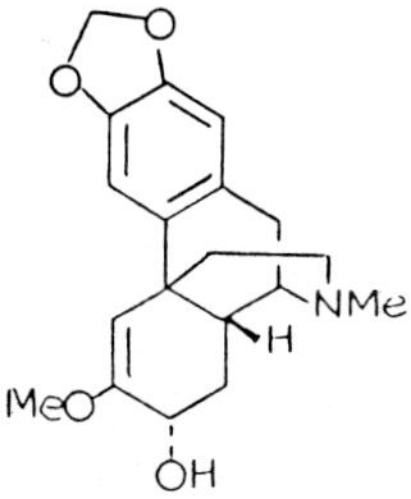

(109)

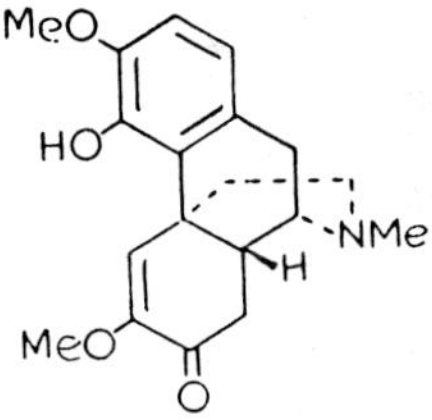

(110)

acutine, from *Ocotea brachybotra* (Vecchietti, Casagrande
and Ferrari, *loc.cit.*)
The structures of these alkaloids have been deduced from their
spectra properties and by their conversion into known bases
and in the cases of sebiferine and tridictopylline the structures were confirmed by X-ray crystallographic studies.

8. *Hasubanan Alkaloids*

New hasubanan alkaloids have been isolated as follows from
        species:
  Oxohasubanonine (III) and oxoprometaphanine (112) from
    S.japonica (A.J. Van Wyk, J.S. African chem.Inst.,
    1975, 28, 284)
  Longanone (113) which is 3-*O*-demethylepistephamiersine,
    and longanine (114) from *S.longa* (A.Lao, Z.Tang and
    R. Xu, Yaoxue Xuebao, 1981, 16, 940; Lao *et al.*, Huaxue
    Xuebao, 1982, 40, 1038)
  Dihydroepistephamiersine (115, R=H) and its acetyl ester
    (115, R=Ac) from *S.abyssinica* (D. Gröger, Planta Med.,
    1975, 28, 269)
  Oxostephamiersine (116) and oxostephasunoline (117) from
    S.japonica (M. Matsui *et al.*, J.nat.Prod., 1984, 47,
    465 and 858)
  Stephavanine dimethyl ether, which is the veratric acid
    ester of *O*-methylstephine (118, R=H), from *S.abyssinica*
    (Van Wyk and A. Wiechers, J.S.African chem.Inst., 1974,
    27, 95)
  Stephabenine, which is the benzoate of *O,N*-dimethylstephine
    (118, R=Me), from *S.japonica* (S.Kondo, Matsui and Y.
    Watanabe, Chem.Pharm.Bull., 1983, 31, 2574)
  Stephadiamine (119) from *S.japonica* (T. Taga, N. Akimoto
    and T. Ibuka, Chem.Pharm.Bull., 1984, 32, 4223)
  1:1'-Bisaknadinine from *S.sasakii* (J. Kunimoto *et al.*,
    Phytochem., 1980, 19, 2735)
Oxohasubanonine is identical with material of structure
(III) previously prepared as an intermediate in the synthesis
of hasubanonine (2nd edn., Vol.IVC p.311) and the other structures were confirmed by chemical correlations with known alkaloids except that the structure of the unusual ring-C contrac-
ted base stephadiamine was determined crystallographically.

9. *Homomorphine Alkaloids*

Two new homomorphine alkaloids collutine (120) (N.L.Mukhamed'-

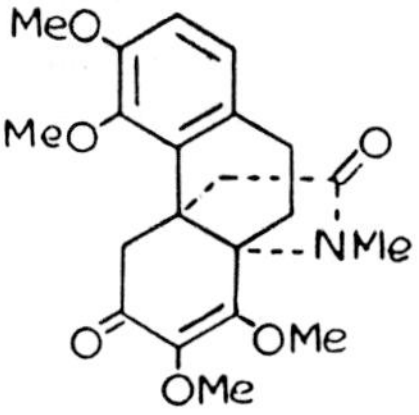

(111)

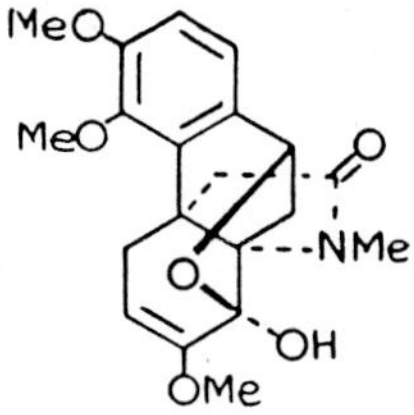

(112)

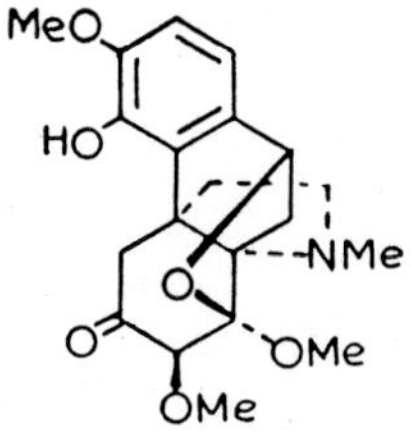

(113)

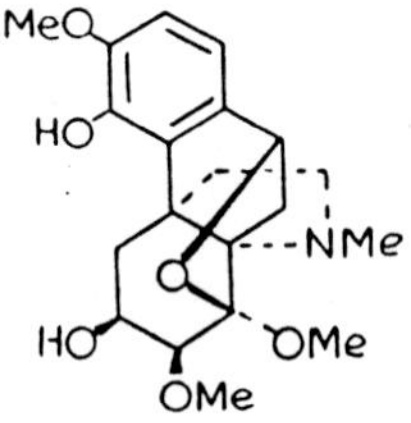

(114)

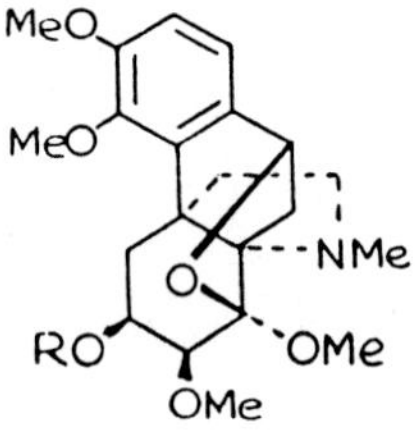

(115)

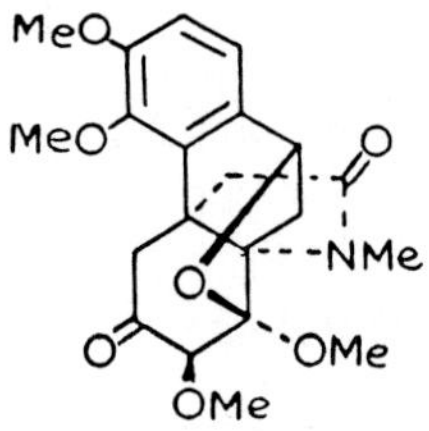

(116)

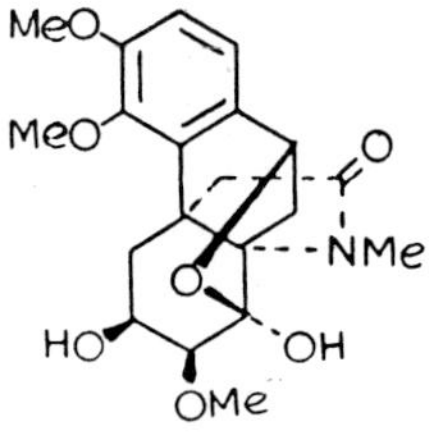

(117)

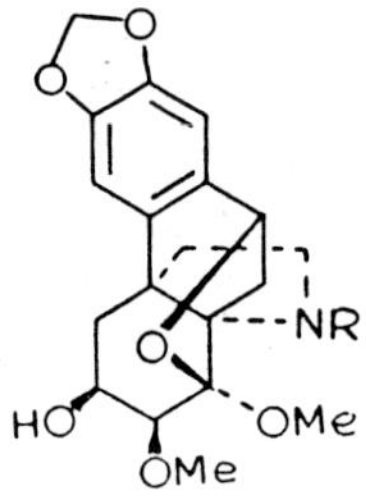

(118)

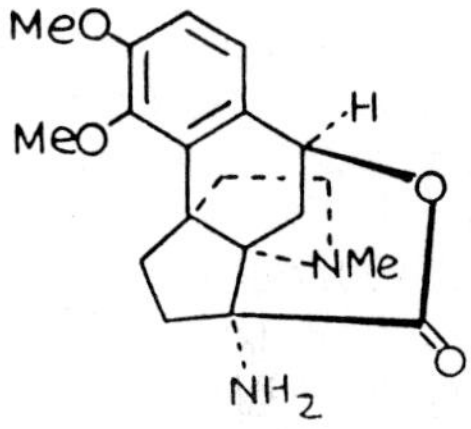

(119)

yarova *et al.*, Khim.Prir.Soedin., 1975, <u>11</u>, 758) and szovitsi-
dine (121) (M.K. Yusupov, Kh.A. Aslanov and Dinh Thi Ngo,
Khim.Prir.Soedin., 1975, <u>11</u>, 271) have been isolated from
Colchicum luteum and *C.szovitsii* respectively.

(120)   (121)

10. Syntheses in the Morphine-Hasubanan Group

Most approaches to the synthesis of alkaloids of the morphine
group have started from benzyltetrahydroisoquinolines. Reduction of the base (122, R^1=OCH$_2$Ph, R^2=CH$_2$Ph) with sodium and
liquid ammonia gives the enol ether (123, R=OH), which can
be hydrolysed and cyclised in acid to 2-hydroxydihydrothebain-
one (124, R=OH). Conversion of this phenol selectively into
the 2-phenyltetrazolyl ether, followed by reduction, yields
dihydrothebainone (124, R=H). This process gives better yields than the earlier route starting from (122, R^1=R^2=H), in
which the intermediate (123, R=H) can cyclise *para* as well
as *ortho* to the hydroxyl group to give 2,3 as well as 3,4-
substituted morphinans (P.R. Crabbendam *et al.*, Recl.J.R.
Neth.chem.Soc., 1983, <u>102</u>, 135).

Morphinandienones can be prepared in very poor yields (<
1%) by Pschorr cyclisation of the diazonium salts obtained
from aminobenzyltetrahydroisoquinolines and also by the cyclisation of bromobenzyltetrahydroisoquinolines photochemically
or by the action of sodamide. For example flavinantine (125,
R=H) has been prepared from (126, R^1=H, R^2=NH$_2$) and *o*-methyl-
flavinantine (125, R=Me) has been obtained from (126, R^1=Me,
R^2=Br). The major productgs of such cyslisations, however
are aporphines (T. Kamatani *et al.*, Chem.Pharm.Bull., 1971,

(122)   (123)   (124)

(125)   (126)   (127)

Anodic oxidation of non-phenolic benzylisoquinolines, however, gives high yields of morphinandienones. In this way
laudanosine (127, R^1=R^2=Me) is converted into O-methylflavin-
antine (125, R=Me) in 52% yield at a platinum anode in acetonitrile and the yield is increased to 63% by the addition
of bis(acetonitrile)palladium(II) chloride (L.L. Miler, F.R.
Sternitz and J.R. Falck, J.Amer.chem.Soc., 1973, 95, 2651).
Even better yields of amurine are obtained by the electrolysis of (127, R^1R^2=CH₂) in tetrafluoroboric acid (T. Kotani
and S. Tobinaga, Tetrahedron Letters, 1973, 4759). This electrochemical cyclisation gives only the flavinantine orientation
of substituents; attempts to induce cyclisation to 3,4-dioxy-
morphinandienones by the use of 6'bromolaudanosine resulted
in formation of (125, R=Me) with loss of bromine and 6'-chloro-
compounds suffered cleavage (Miller et al., J.org.Chem., 1978,
43, 1580).

Chemical oxidative coupling of diphenolic benzylisoquinolines affords morphinandienones but the yields are very low
using tertiary bases, owing to the formation of dibenzyopyrr-
ocolines. Non-basic amides, however, give moderate yields of
the desired coupling products, for example N-ethoxycarbonyl-
norreticuline (128) gives 23% of N-ethoxycarbonylnorsalutarid-
ine (129) when oxidised with thallium (III) trifluoroacetate
(M.A. Schwartz and I.S. Mami, J.Amer.chem.Soc., 1975, 97,
1239) and 3-oxoreticuline (130) with iodosobenzene diacetate
in trifluoroacetic acid gives 16-oxosalutaridine (131, R^1=H,
R^2=OH) 25% and 16-oxopallidine (131, R^1=OH, R^2=H) 8% together
with the isopavine oxothalidine (132) (10%) and the aporphine
oxoboldine (6%) whereas the same amide is oxidised by vanadium
oxychloride to oxopallidine (25%), oxothalidine (9%) and oxoboldine (11%) (D.G. Vanderlaan and Schwartz, J.org.Chem.,
1985, 50, 743). The dimethyl ether of (130), however, is
oxidised by vanadium oxychloride and electrolytically to the
proerythradienone (133), which is an analogue of the dienone
into which thebaine is converted by strong acids (I.W. Elliott,
J.org.Chem., 1977, 42, 1090).

(128)   (129)   (130)

(131)   (132)   (133)

8-Chloro-*O*-methylflavinantine (136, R¹=R²=Me) and 8-chloro-amurine (136, R¹R²=CH₂) have been prepared by the cyclisation of the acetoxydienones (134) (obtained by oxidising the appropriate benzyltetrahydroisoquinolines with lead tetra-acetate) in trifluoroacetic acid. Other products of these cyclisations are aporphines and isopavines (H. Hara, O. Hoshino and B. Umezawa, *Heterocycles*, 1976, 5, 1802).

(134)   (135)   (136)

Two novel syntheses, which do not proceed through benzyl-isoquinolines, have been developed. In the first of these the tetrahydropyridine derivative (137) is alkylated, *via* the mesomeric anion (138) with 2-bromoethyl-4-bromobutene to give the enamine (139), which is readily cyclised to the iminium salt (140). Treatment of the iminium salt with diazomethane gives the aziridinium salt (141), which is oxidised with di-methylsulphoxide to the aldehyde (142). The aldehyde is cyclised by boron trifluoride to the alcohol (143), which is unable to lose water for steric reasons. The hydroxyl group is removed by the reduction of its methanesulphonyl ester and the resulting olefin is converted into the diol (144). Cleavage of this diol gives *O*-methyl-β-dihydrothebainone (145, R=Me), which is demethylated to β-dihydrothebainone (145, R=H). The latter is an intermediate in Gates's first synthesis of morphine so the route constitutes a formal synthesis of the alkaloid (D.A. Evans and C.H. Mitch, *Tetrahedron Letters*, 1982, 23, 285; W.H. Moos, R.D. Gless and H. Rapoport, *J.org. Chem.*, 1983, 48, 227).

The 2,3-dimethoxy isomers (146) of *O*-methyl-β-dihydratheb-

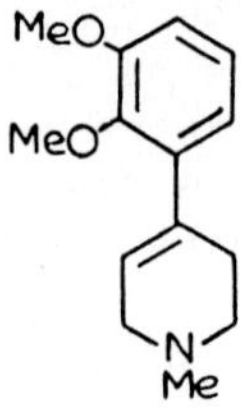

(137)

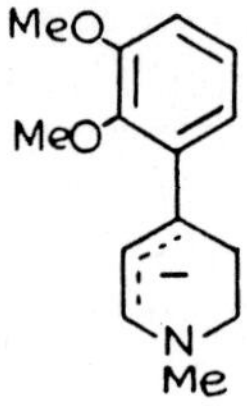

(138)

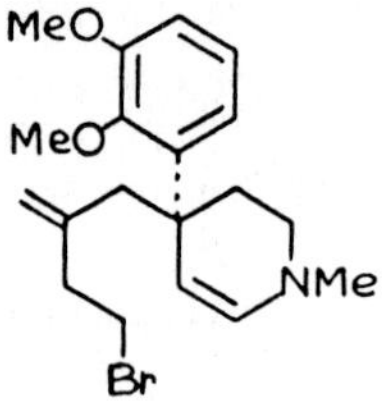

(139)

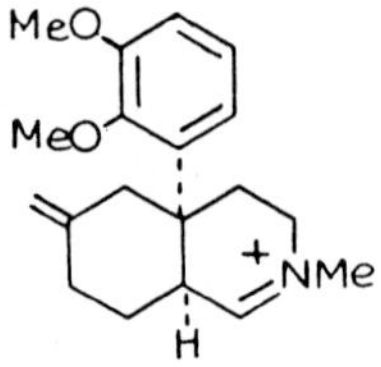

(140)

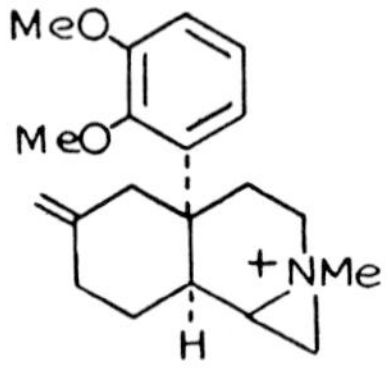

(141)

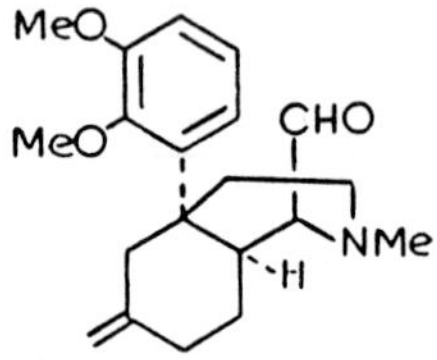

(142)

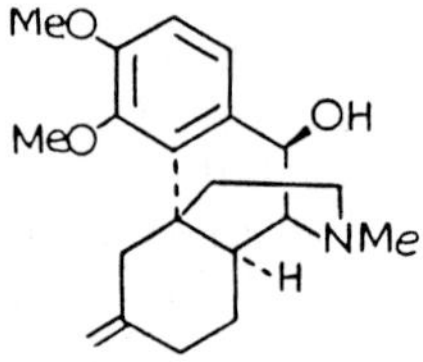

(143)

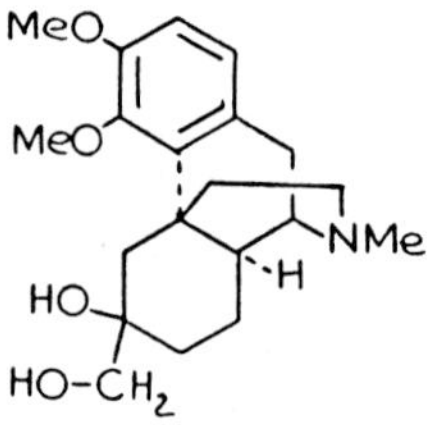

(144)

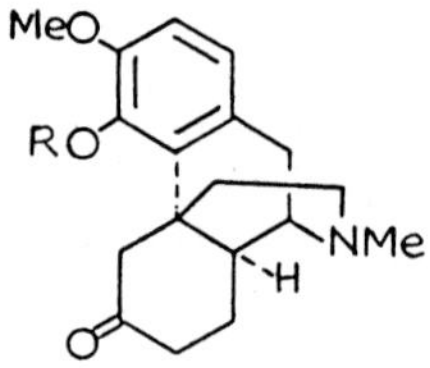

(145)

ainone has been prepared by a similar route from the corresponding isomer of (137) and converted through the enamine (147) into the dithioketal (148), hydrolysis of which affords the α-diketone (149). Conversion of this diketone into its enol methyl ether by treatment with methanol and toluenesulphonic acid gives a 4:1 mixture of O-methylpallidinine (150) and its 6-keto-7-methoxy isomer (151) (J.E. McMurry *et al.*, J.org.Chem., 1984, 49, 3803).

(146)   (147)   (148)

(149)   (150)   (151)

The other novel synthesis also involves the closure of
ring B at a late stage in the process. The starting material
is the complex isoquinolone derivative (152), the photolysis
of which affords the benzodihydrofuran (153, R¹=Ac, R²=H),
which is converted into (153, R¹=R²=Me). Ketalisation of
this, followed by hydrolysis and decarboxylation gives the
lactam (154), as a mixture of the *cis* and the *trans* isomer,
reduction of which with di-isobutylaluminium hydride yields
the enamine (155). Protonation of the enamine occurs on the
least hindered side to give the inimium salt which reacts
with potassium cyanide to give the nitrile (156). Treatment
of the nitrile with methyl lithium, followed by mild hydrolysis of the resulting imine, gives the keto-ketal (157), cyclodehydration of which in acid is accompanied by hydrolysis
to the keto-olefin (158). Ketalisation and ozonolysis of this
gave the ketone (159), which is reduced to B/C-*trans* dihydrocodeinone dimethyl ketal (160) (A.G. Schultz *et al.*, J.org.
Chem., 1985, 50, 217).

(152)   (153)   (154)

(155)   (156)   (157)

(158)   (159)   (160)

In the hasubanan group of alkaloids metaphanine (166) has
been synthesised from an intermediate in a previously reported
synthesis of hasubanonine. This compound (161) is converted
into the ketal and oxidised to the C-10 ketone (162, $R^1=R^2=Ac$),
which is hydrolysed and methylated in alkali to (162, $R^1=Me$,
$R^2=H$) and then reduced to the diol (163). The less hindered
hydroxyl group of the diol is protected as its tetrahydropyr-

anyl ether to allow the oxidation at C-8 to the ketone, after which removal of the protecting group gives the hemi-acetal (164). Selective reduction of the amide is achieved by treatment with boron trifluoride followed by sodium borohydride to give the ketal (165), which is hydrolysed to metaphanine (166). (T. Ibuka, K. Tanaka and Y. Inubushi, Chem.Pharm.Bull., 1974, 22, 907).

(161)   (162)   (163)

(164)   (165)   (166)

A synthesis of (±)-cepharamine has been accomplished from reticuline (167) by degradation to (168) and reduction to (169), which when oxidised by vanadium oxychloride or, better, by titanium trifluoroacetate, gives a mixture of (170, R¹=H, R²=OH) and (170, R¹=OH, R²=H). Hydrolysis of the amide is accompanied by addition of the secondary amine to the unsaturated ketone and (170, R¹=H, R²=OH) gives (171), which is converted by methanolic hydrogen chloride into the isomeric enol ether cepharamine (172) (Kametani *et al.*, Tetrahedron, 1974, 30, 1059; Schwartz and R.A. Wallace, Tetrahedron Letters, 1979, 3257).

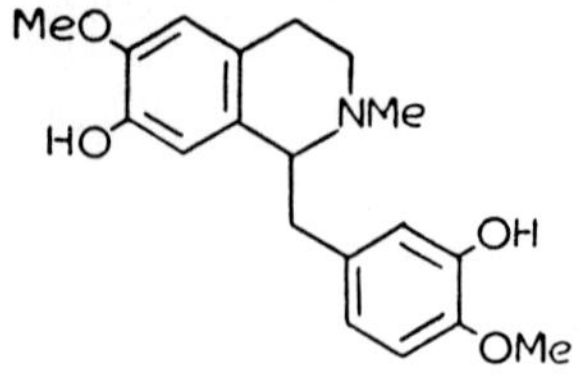

(167)

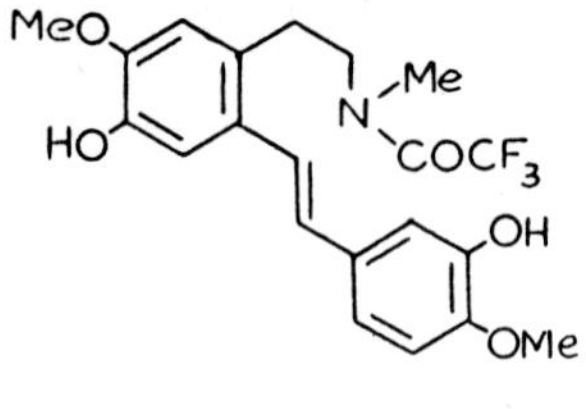

(168)

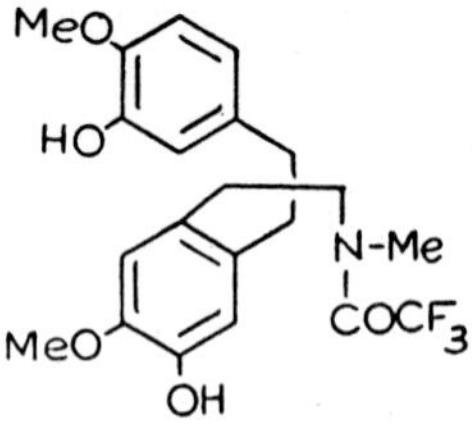

(169)

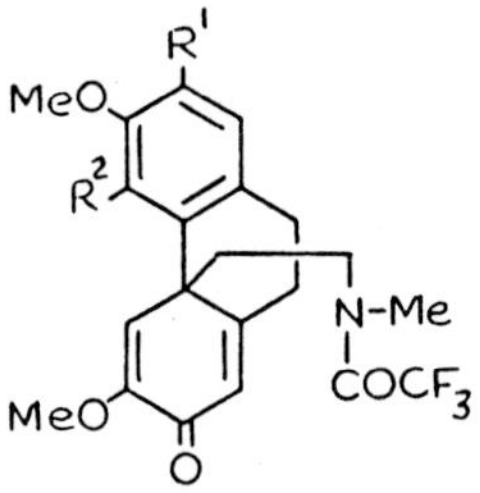

(170)

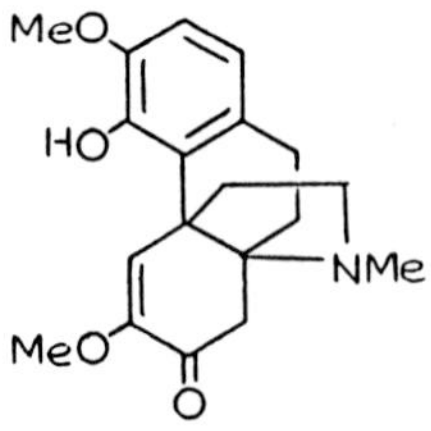

(171)

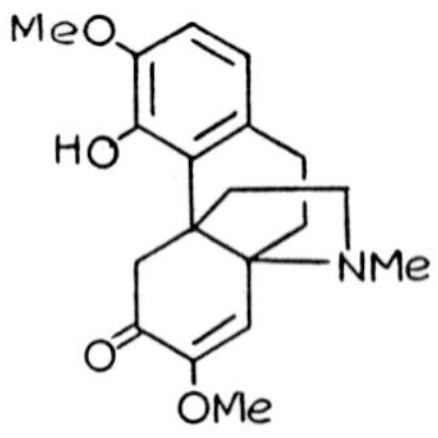

(172)

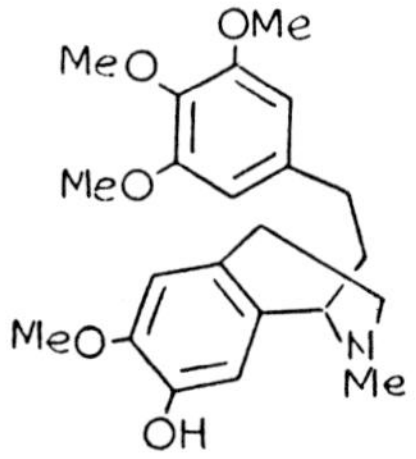

(173)

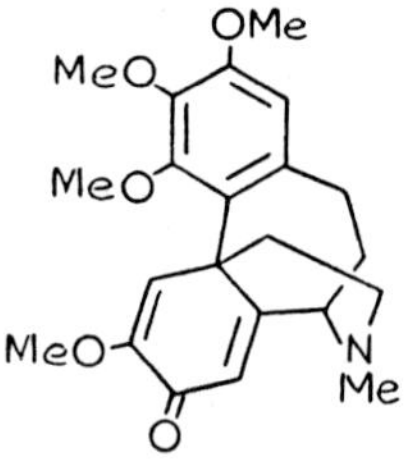

(174)

Oxidative coupling by thallium trifluoroacetate has also been used for the preparation of the homomorphinandienone (±)-*O*-methylandrocymbine (174) from the phenethylisoquinoline *O*-methylautumnaline (173), the production of unwanted dibenzo-quinolizinium salts being inhibited by complexing the tertiary nitrogen with diborane (Schwarz and B.F. Rose, J.Amer.chem. Soc., 1973, <u>95</u>, 612).

Chapter 37

FUSED HETEROCYCLIC SYSTEMS HAVING A NITROGEN ATOM COMMON TO
TWO OR MORE RINGS

D. LEAVER

1. *Indolizines*

The chemistry of indolizines, including that of hydrogenated
derivatives, has been reviewed (F.J. Swinbourne, J.H. Hunt,
and G. Klinkert, Adv. Heterocyclic Chem., 1978, *23*, 103; W.
Flitsch in 'Comprehensive Heterocyclic Chemistry', ed. A.R.
Katritzky and C.W. Rees, Pergamon Press, Oxford, 1984, vol.
4, p. 443).

(a) *Aromatic indolizines*

Methods for the synthesis of aromatic indolizines are the
subject of two reviews (N.S. Prostakov and O.B. Baktibaev,
Russ. chem. Rev., 1975, *44*, 748; T. Uchida and K. Matsumoto,
Synthesis, 1976, 209).
    Indolizines are commonly prepared from quaternary pyri-
dinium salts by methods involving either carbonyl condensation reactions, as in the very important Tschitschibabin
reaction (2nd. edn. Vol. IVH, p. 259), or pericyclic
reactions (1,3-dipolar cycloaddition or 1,5-electro-
cyclisation) *via* the pyridinium ylides.
    In its original form, the cycloaddition procedure involves
the formation of an acyl-stabilised pyridinium ylide (e.g. 1)

and its reaction, *in situ*, with an acetylenic dipolarophile
such as dimethyl acetylenedicarboxylate (DMAD).   Dehydrogenation of the initial adduct is carried out *in situ* with
a palladium catalyst (V. Boekelheide and K. Fahrenholz, J.
Amer. chem. Soc., 1961, *83*, 458) or it occurs spontaneously
(C.A. Hendrick, E. Ritchie, and W.C. Taylor, Austral. J.
Chem., 1967, *20*, 2467).   The need for a dehydrogenation
step may be avoided by using a sulphonyl-stabilised
pyridinium ylide (2), when aromatisation occurs by loss of
p-toluenesulphinic acid and the yield (of 3) is much improved
(R.A. Abramovitch and V. Alexanian, J. org. Chem., 1976, *41*,
2144).

(2)   (3)

Unstabilised pyridinium ylides, generated by fluoride-
induced desilylation of *N*-(trimethylsilylmethyl)pyridinium
salts, react with DMAD to give the same products (3) (O.
Tsuge *et.al.*, Chem. Letters, 1984, 281;   Y. Miki *et.al.*,
Heterocycles, 1984, *22*, 701).

1,3-Cycloaddition reactions of pyridinium ylides can also
be carried out with olefinic dipolarophiles, a notable
example of which is phenyl vinyl sulphoxide.   This compound
reacts with pyridinium dicyanomethylide (4) to give 3-cyano-
indolizine (5, 43%), m.p. 46-48°C, aromatisation of the
initial adduct being brought about by loss of hydrogen
cyanide and benzenesulphenic acid (K. Matsumoto, T. Uchida,
and L.A. Paquette, Synthesis, 1979, 746).

(4)   (5)

The 1,5-cyclisation reaction of pyridinium ylides (Y. Tamura *et.al.*, J. chem. Soc. Perkin I, 1973, 2091; T. Sasaki *et.al.*, *ibid.*, p. 2089) is exemplified in the formation (45%) of the diester (3) when the quaternary salt (6) is treated with potassium carbonate.

For the preparation of indolizine itself, dehydrogenation of 3-(2-pyridyl)propanol (2nd. edn. Vol IVH, p. 260) is probably still the best route. This reaction is believed to proceed *via* the corresponding pyridylpropanal which cyclises to indolizine with loss of water, as has been observed for more highly substituted 2-(3-oxoalkyl)pyridines. A related process is involved in the synthesis (70%) of 8-acetoxyindolizine (9), b.p. 82°C/0.1 mm., from the Mannich base (7) and an alkyl vinyl ether; the intermediate product (8) is a cyclic acetal of 3-(3-hydroxy-2-pyridyl)propanal (D. Blondeau and H. Sliwa, J. chem. Res., 1979 (S) 2, (M) 117).

Although most of the important syntheses of indolizines proceed by building the pyrrole ring on to a preformed pyridine ring, there is a noteworthy example of the reverse procedure in the preparation (80%) of 8-cyanoindolizine (12), lemon-yellow needles, m.p. 74-75°C, from 2-pyrroleacetonitrile (11) and the vinamidinium salt (10) (C. Jutz, R.M. Wagner, and H-G. Löbering, Angew. Chem. intern. Edn., 1974, *13*, 737).

Indolizine and its alkyl derivatives are easily oxidised,

(10)   (11)   (12)

even by exposure to air.   The products of oxidation are
often highly coloured and ill-defined but, by the action of
a palladium-charcoal catalyst (A. Kakehi *et.al.*, Bull. chem.
Soc. Japan, 1981, *54*, 2833) or potassium hexacyanoferrate
(III) (S. Hünig and F. Linhart, Tetrahedron Letters, 1971,
1273), 3,3'-biindolizines (13) may be obtained.   This oxidative dimerisation is believed to proceed *via* indolizine
radical-cations.

(13)

(b) *Indolizidines and partially reduced indolizines*

The indolizidine ring system is present in many alkaloids of
diverse origins and this subject is reviewed regularly (see
M.F. Grundon, Nat. Prod. Reports, 1985, *2*, 235).

   Among the more novel routes to indolizidines is the transannular cyclisation of azonines (S.R. Wilson and R.A.
Sawicki, J. org. Chem., 1979, *44*, 330).   Beckmann rearrangement of the cyclo-octenone oxime (14) gives a mixture of
hexahydroazoninones (15) and (16) which, upon successive
treatment with mercury(II) acetate and sodium borohydride
yields (85%) 5-oxoindolizidine (17), ν_{max} 1640 cm^{-1}.   Reduction with lithium aluminium hydride then yields indolizidine (82%).

(14)   (15)   (16)   (17)

The indolizidine ring system may be constructed in one
step from an acyclic precursor by making use of an intramolecular Diels-Alder reaction.   There are two variants of
this strategy whereby the nitrogen atom may be a component
either of the dienophile unit (S.M. Weinreb, Acc. chem. Res.,
1985, *18*, 16) or of the diene unit (Y-S. Cheng, A.T. Lupo,
and F.W. Fowler, J. Amer. chem. Soc., 1983, *105*, 7696).
Scheme 1 illustrates the use of these two variants for the
preparation of 3-oxoindolizidine (18), ν_{max} 1657 cm^{-1}, and
shows how, in each case, the precursor molecule is transiently
generated by short contact pyrolysis of an acetate ester.

Scheme 1

Bridge-bond iminium salts such as 2,3,5,6,7,8-hexahydro-
1*H*-indolizinium perchlorate (19), m.p. 218-219°C decomp., are
important in the chemistry of indolizidines and other 1-aza-.

bicyclo[n,m,O]alkanes because they permit the attachment of
nucleophile-derived groups to the bridgehead carbon atom.
Such iminium salts are the conjugate acids of enamines (20)
(review by O. Cervinka in 'Enamines', ed. A.G. Cook, Marcel
Dekker, New York, 1969, p. 253) and have traditionally been
obtained from the parent azabicycloalkanes by dehydrogenation
with mercury(II) acetate.

(19)   (20)

A new and very simple route to these iminium salts is pro-
vided by heating N-(carboxyalkyl)lactams (21) with soda-lime
and treatment of the resulting enamines with perchloric acid
(S. Miyano *et.al.*, J. heterocyclic Chem., 1982, *19*, 1465):

(21)

Indolizine derivatives hydrogenated only in the pyridine
ring are accessible by cyclisation of *N*-substituted pyrroles.
Thus the preparation (59%) of 5,6,7,8-tetrahydroindolizin-8-
one (22), faintly yellow prisms, m.p. 34°C, from *N*-(3-cyano-
propyl)pyrrole by a modified Hoesch reaction, has long been
known (J.M. Patterson, J. Brasch, and P. Drenchko, J. org.
Chem., 1962, *27*, 1652).   A more recent example is the
preparation (100%) of the isomeric 7-oxo-compound (23), a
colourless oil, by copper-catalysed thermolysis of the diazoketone (24) (C.W. Jefford and W. Johncock, Helv., 1983, *66*,
2666).

(22)   (23)   (24)

2. *Quinolizines*

(a) *Aromatic quinolizines*

Reviews: G. Jones, Adv. heterocyclic Chem., 1982, *31*, 2;
C.K. Bradsher in 'Comprehensive Heterocyclic Chemistry', ed.
A.R. Katritzky and C.W. Rees, Pergamon Press, Oxford, 1984,
Vol. 2, p. 525; S-D. Saraf, Heterocycles, 1980, *14*, 2047
(acridizinium salts).

The method of choice for the preparation of salts (1) of
the parent quinolizinium ion is the dehydration of 1-oxo-
1,2,3,4-tetrahydroquinolizinium bromide (2), m.p. 198°C
(picrate, m.p. 161-162°C) with acetic anhydride (E.E. Glover
and G. Jones, J. chem. Soc., 1958, 3021; T. Miyadera and I.
Iwai, Chem. pharm. Bull., 1964, *12*, 1338).

(1)   (2)

The oxo-compound (2) is also a key intermediate for the preparation of 1-hydroxy- and 2-bromoquinolizinium salts (A.
Fozard and G. Jones, J. chem. Soc., 1963, 2203) and, *via* its
oxime, for that of the 1-amino-compound (A.R. Collicutt and
G. Jones, J. chem. Soc., 1960, 4101).   3-Bromoquinolizinium
bromide is obtained by dehydration of the 7-bromo-derivative
of (2) and 1-bromoquinolizinium bromide by heating quinolizinium perbromide (1; X=Br$_3$) at 200°C (G.M. Sanders, M. van
Dijk, and H.C. van der Plas, Heterocycles, 1981, *15*, 213).
4-Halogenoquinolizinium salts are obtained by treatment of
4*H*-quinolizin-4-one (2nd. edn. Vol. IVH, p. 265) with phosphoryl halides (J.A. van Allen and G.A. Reynolds, J. org.
Chem., 1963, *28*, 1022; G.M. Sanders *et.al.*, *loc.cit.*).

Other major routes to quinolizinium salts involve aldol-
type condensations in quaternary pyridinium salts.   These
reactions may be intramolecular, as in the preparation of 3-
hydroxyquinolizinium bromide by heating the salt (3) with 50%
aqueous hydrogen bromide (P.A. Duke, A. Fozard, and G. Jones,
J. org. Chem., 1965, *30*, 526), or intermolecular, as in the
base-catalysed reactions of α-diketones with salts of the
general type (4) (O. Westphal *et.al.*, Arch. Pharm., 1961,

294, 37; T.L. Hough and G. Jones, J. chem. Soc. (C), 1968,
1082; J. Alvarez-Builla *et.al.*, J. heterocyclic Chem.,
1985, *22*, 681).

(3)   (4)   (5)

The latter synthesis generally leads to 2,3-disubstituted
quinolizinium salts (5), the ethoxycarbonyl group being lost
during the condensation.
     The most important reactions of quinolizinium salts are
nucleophilic addition and substitution.   The parent ion (1)
reacts with nucleophiles at C-4 and the resulting transient
4*H*-quinolizines (6) undergo rapid ring-opening to form 1-(2-
pyridyl)butadienes (7).   Such reactions have been observed

(6)   (7); Nu = H, alkyl,
                                aryl, or NR$_2$

with lithium aluminium hydride and with Grignard reagents
(T. Miyadera *et.al.*, Chem. pharm. Bull., 1964, *12*, 1344;
Tetrahedron, 1969, *25*, 397) and with secondary amines (D.
Mörler and F. Kröhnke, Ann., 1971, *744*, 65).   When 1-amino-
quinolizinium salts are diazotised in aqueous solution, the
exceptionally electrophilic diazonium dications (8) react in
this way with water and the resulting ring-opened diazonium
ions (9) recyclise to 1,2,3-triazolo[1,5-*a*]pyridines (10)
(L.S. Davies and G. Jones, J. chem. Soc (C), 1970, 688).
     Nucleophilic displacement of halide ion occurs in 2- and
4-halogenoquinolizinium salts but halogens in the 1- and 3-
positions are inert to substitution (J.A. van Allen and G.A.

(8)   (9)   (10)

Reynolds, *loc.cit.*; G.M. Sanders *et.al.*, J. heterocyclic Chem., 1982, *19*, 797).

4*H*-Quinolizines (6) and 1-(2-pyridyl)butadienes (7) are partners in a ring-chain tautomeric system but, in the absence of ring-stabilising substituents, only the chain tautomer has been observed directly. The ring tautomer is stabilised by electron-withdrawing groups at C-1 and C-3 (quinolizine numbering) and many examples of such stable 4*H*-quinolizines are known (2nd. edn. Vol. IVH, p. 266). However, there is only one known example, shown below, containing a single ring-stabilising substituent at C-3, in which both tautomers can be detected (by [1]H-nmr spectroscopy) at equilibrium (G.G. Abbot, D. Leaver, and K.C. Mathur, J. chem. Res., 1978, (S) 224, (M) 2850).

Whereas the conjugate bases of 2- and 4- hydroxy-quinolizinium ions are usually represented as 2- and 4-quinolizones, (11) and (12) (2nd. edn. Vol. IVH, p. 265), those derived from 1- and 3-hydroxyquinolizinium ions have no corresponding uncharged canonical structures and are represented as betaines, (13) and (14).
The betaines (14) react with acetylenic dipolarophiles at C-4 and C-6 to give, after dehydrogenation, [2.3.3]cycla-zinones [see section 4(b)].

(11)　　　　　　(12)　　　　　　(13)　　　　　　(14)

(b) *Quinolizidines and partially reduced quinolizines*

Conformational aspects of quinolizidine chemistry are included in a review by T.A. Crabb and A.R. Katritzky (Adv.
heterocyclic Chem., 1984, *36*, 1).
　　The chemistry of quinolizidines has much in common with
that of indolizidines [section 1(b)].　　Thus the intramolecular Diels-Alder reaction is equally applicable in the
quinolizidine series, as exemplified in the synthesis of the
hexahydroquinolizinone (15) (Y-S. Cheng *et.al.*, *loc.cit.*)
and heating the *N*-(carboxyalkyl)lactam (16) with soda-lime
provides a convenient route to the enamine (17) and its
derived iminium salt (18) (J.M. McIntosh, Canad. J. Chem.,
1980, *58*, 2604).

(15)　　　　　　(16)　　　　　　(17)　　　　　　(18)

3.　*Azepine derivatives containing a bridgehead nitrogen atom*

Pyrrolo[1,2-*a*]azepines are included in a review on the
chemistry of aza-azulenes (T. Nishiwaki and N. Abe, Heterocycles, 1981, *15*, 547).
　　In addition to 5*H*-pyrrolo[1,2-*a*]azepin-5-one (1) (2nd.
edn. Vol. IVH, p. 270), the two isomeric oxo-compounds, (2)
and (3) are now known.　　9*H*-Pyrrolo[1,2-*a*]azepin-9-one (3),
m.p. 106-107°C, ν_{max} 1650 cm^{-1}, first known in the form of its

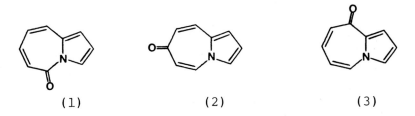

(1)   (2)   (3)

1,2,3-tribromo-derivative (E.W. Collington and G. Jones,
J. chem. Soc. (C), 1969, 1028), is prepared from 2-acetyl-
pyrrole and the vinamidinium salt (4) (W. Flitsch, F.
Kappenburg, and H. Schmitt, Ber., 1978, *111*,, 2407).

Me$_2$N$^+$ + CH$_3$ HN — NaOMe → HN NMe$_2$ — 400°C, 0.1 torr → (3)

ClO$_4^-$

NMe$_2$

(4)

7*H*-Pyrrolo[1,2-*a*]azepin-7-one (2), yellow needles, m.p. 119-
120°C, ν_{max} 1650 cm^{-1}, is prepared from pyrrole-2-carbalde-
hyde and but-3-en-2-one (G. Jones and P.M. Radley, J. chem.
Soc., Perkin I, 1982, 1123).

O= CH$_3$ + H HN — 1) OH$^-$ 2) OMe$^-$ → O= N — Pd-C, 170°C → (2)

(5)

Nmr studies show that, in strongly acidic solutions, all
of the pyrroloazepinones (1)-(3) are protonated on the oxygen
atom, the highly coloured conjugate acids being hydroxy-
derivatives of the as yet unknown 3a-azonia-azulenium ion (6).
An unsubstituted pyrrolo[1,2-*a*]azepine, the 5*H*-isomer (7) is
obtained (31%) from the ketone (5) by reduction to the alcohol
and dehydration; it is not converted, by hydride abstraction,

(6)   (7)

into the ion (6) (G. Jones and P.M. Radley, *loc.cit.*).
Another tautomeric form of pyrrolo[1,2-*a*]azepine is formed
(56%) as a tricyclic derivative, 10*H*-azepino[1,2-*a*]indole
(9), m.p. 91.5°C, when *o*-azidodiphenylmethane (8) is heated
at 200°C in 1,2,4-trichlorobenzene (G.R. Cliff, E.W. Collington, and G. Jones, J. chem. Soc. (C), 1970, 1490).   This
reaction is believed to proceed *via* an intermediate aryl-
nitrene (ArN:) which inserts intramolecularly into the unsubstituted benzene ring.

(8)   (9)

A red compound, previously regarded as a derivative (10)
of pyrido[1,2-*a*]azepine (2nd. edn. Vol. IVH, p. 271), is now
reformulated as the monocyclic compound (11) (G. Jones and
P. Rafferty, J. org. Chem., 1982, *47*, 2792).   This ring
system is therefore unknown in its fully unsaturated form.

(10)   (11)

Partially saturated [*a*]-fused azepines (14) may be synthesised by heating cyclic α-aminoacids or their acyl derivatives (12), in acetic anhydride, with 1,2-dicyanocyclobu-
tene (I.J. Turchi, C.A. Maryanoff, and A.R. Mastrocola, J.
heterocyclic Chem., 1980, *17*, 1593).   The reaction is
believed to proceed by 1,3-dipolar cycloaddition of the

cyclobutene double bond to a mesoionic intermediate (13) followed by loss of carbon dioxide and opening of the cyclo- butane ring (Scheme 2).

(12); n=1,2, or 3

−H₂O

(13)

−CO₂

(14)

Scheme 2

4. *Tricyclic compounds with nitrogen common to all three rings*

The chemistry of cyclazines is the subject of two reviews (A. Taurins in 'Special Topics in Heterocyclic Chemistry', ed. A. Weissberger and E.C. Taylor, Wiley-Interscience, New York, 1977, p. 245; W. Flitsch and U. Krämer, Adv. hetero- cyclic Chem., 1978, *22*, 321) and is included in a chapter of 'Comprehensive Heterocyclic Chemistry' (W. Flitsch, Vol. 4, p. 478).

A proposal to modify the nomenclature of cyclazines by placing the bracketed numerals, in increasing order, before the word 'cyclazine' rather than in the middle of it (M.A. Jessep and D. Leaver, J. chem. Soc. Perkin I, 1980, 1319) has been widely adopted.

(a) *[2.2.3]Cyclazines (pyrrolo[2,1,5-cd]indolizines)*

These compounds (3) may be prepared from 3*H*-pyrrolizine (1; R=H) and its derivatives (R=Ph, CO₂Me, SO₂Ph) by heating with the vinamidinium salt (2) and sodium hydride in *N*,*N*- dimethylformamide (V. Batroff *et.al.*, Ber., 1984, *117*, 1649).

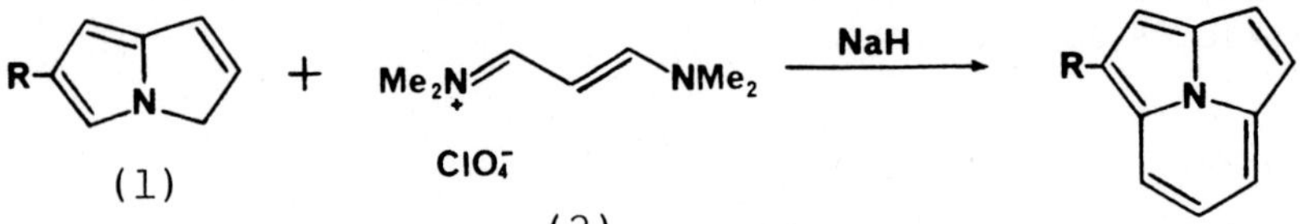

(1)   (2)   (3)

1,3,4,4a,5,6,7,7a-Octahydro[2.2.3]cyclazine (4), m.p. 93°C,
perchlorate, m.p. 172°C, is obtained by a method similar to
the *N*-(carboxyalkyl)lactam cyclisation described in sections
1(b) and 2(b).   It can be dehydrogenated to [2.2.3]cyclazine
(24%) with palladium-charcoal at 200°C (K. Takada, J. Haginiwa,
and I. Murakoshi, Chem. pharm. Bull., 1976, *24*, 2265).

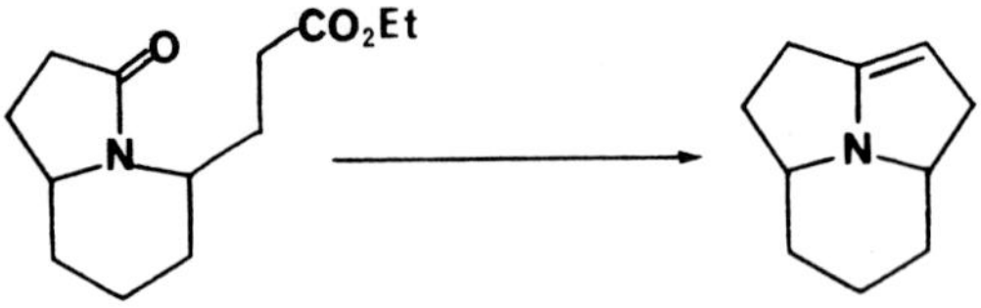

(4)

(b) [2.3.3]*Cyclazines* (*pyrrolo*[2,1,5-de]*quinolizines*) *and*
     [2.2.4]*cyclazines* (*azepino*[2,1,7-cd]*pyrrolizines*)

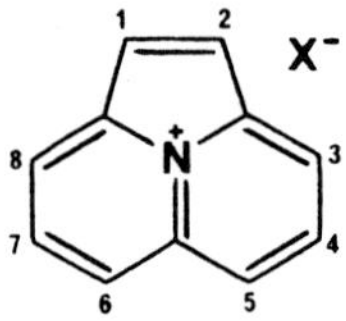

(5)

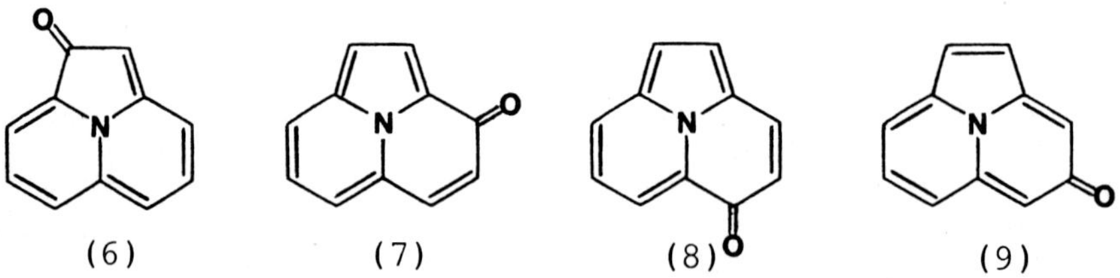

(6)   (7)   (8)   (9)

The parent aromatic molecule of the [2.3.3]cyclazine series
is a cation (5), the [2.3.3]cyclazinylium ion.   Various
isomeric [2.3.3]cyclazinones (6)-(9) are also possible and
the first three of these are known (D. Farquhar $et.al.$, J.
chem. Soc. Perkin I, 1984, 2553).

$1H$-[2.3.3]Cyclazin-1-one (6) is obtained in a partially
hydrated form, red needles, m.p. 186-192°C, ν_{max} 1620 cm^{-1},
by thermal cyclisation of the quinolizine derivative (10)
followed by hydrolysis and decarboxylation of the resulting
oxo-ester (11).   Treatment of the cyclazinone (6) with

phosphorus pentasulphide yields an unstable thione which is
converted into a [2.3.3]cyclazinylium salt (5) by methylation
(on sulphur) and hydrogenolysis with Raney nickel. $3H$-
[2.3.3]Cyclazin-3-one (7), deep yellow needles, m.p. 165-
167°C, ν_{max} 1590 cm^{-1}, is prepared from the conjugate base
(12) of the 3-hydroxyquinolizinium ion [section 2(a)] by reaction with ethyl propynoate in boiling nitrobenzene, followed
by hydrolysis and decarboxylation of the oxo-ester (13).

Treatment of the $3H$-cyclazin-3-one (7) with phosphoryl
bromide yields a 3-bromocyclazinylium salt which is converted,
by catalytic hydrogenolysis, into a [2.3.3]cyclazinylium salt
identical with the product obtained from the $1H$-cyclazin-
1-one.

Various 2,4-disubstituted $3H$-[2.3.3]cyclazin-3-ones (14)
are obtained from indolizines by intramolecular condensation
reactions as exemplified below (J.W. Dick $et.al.$, J. chem.

Soc. Perkin I, 1981, 3150).

(14)

[2.3.3]Cyclazinylium perchlorate (5; X=ClO₄), needles, m.p. 284-285°C, does not show the characteristic yellow colour of its hydrocarbon analogue, acenaphthylene. Its ^{1}H nmr spectrum shows strong deshielding [δ 8.64 (H-1 and 2), 9.1-9.4 (H-3,4,5,6,7 and 8)] indicative of a ring current pathway that includes the whole periphery of the molecule. The 1,2-bond, unlike that of acenaphthylene, is inert to catalytic hydrogenation at atmospheric pressure and does not act as a dienophile. Nucleophilic attack takes place at C-3 and C-5 as, for example, in the reaction with sodium sulphide which yields a mixture of thiones, oxidised in the presence of air and light to the corresponding cyclazinones (7) and 5H-[2.3.3]cyclazin-5-one (8), a deep yellow solid, m.p. 180-182°C.

[2.2.4]Cyclazinylium perchlorate (16; R=H) is obtained as a red solid (mixed with NaClO₄), δ_H 8.8-9.6, when the pyrrolo-[1,2-a]azepine (15; R=CO₂CMe₃) is treated with N,N-dimethyl-formamide and phosphoryl chloride. The 1-cyano-compound (16; R=CN) is prepared similarly from (15; R=CN), and other [2.2.4]cyclazinylium salts are formed by reaction of the pyrrolizine derivative (17) with an enol ether or an enamine (W. Flitsch and E.R.F. Gesing, Ber., 1983, *116*, 1174).

(15)

(16)

(17)

(c) [3.3.3]*Cyclazines* (*pyrido*[2,1,6-de]*quinolizines and* [2.3.4]*cyclazines* (*azepino*[2,1,7-cd]*indolizines*)

[3.3.3]Cyclazine (18) has an unusually low first ionisation

potential (5.9 eV) consistent with the ready formation of
its radical-cation salts (M.H. Palmer *et.al.*, J. mol.
Structure, 1977, *42*, 85).   Its first electronic transition,
which gives rise to an absorption band in the near infrared
(1290 nm), is also of unusually low energy (W. Leupin and J.
Wirz, J. Amer. chem. Soc., 1980, *102*, 6068).   A new synthesis
of the 1,3-di(ethoxycarbonyl)-derivative (19) gives an improved yield (50-60%) in essentially one operational step from
4-chloroquinolizinium perchlorate and diethyl glutaconate
(D. Leaver, Pure and appl. Chem., 1986, *58*, 143).

(18)   (19)

Diethyl [2.3.4]cyclazine-4,5-dicarboxylate (22), black-
violet needles, m.p. 87°C, δ_H 3,50-4.55 (azepine H), 5.13,
5.25 and 6.10 (indolizine H), is prepared from the pyrrolo-
[1,2-*a*]azepinone (20) and the Wittig reagent (21) by a two-
stage condensation.   It reacts with *N*-phenylmaleimide in a
Diels-Alder cycloaddition across the 6- and 9-positions and
is converted into the 6,7,8,9-tetrahydro-derivative under mild
conditions (W. Flitsch *et.al.*, Ber., 1975, *108*, 2969;   1979,
112, 3577).

(20)   (21)   (22)

In its fully hydrogenated (perhydro) form, [3.3.3]cyclazine
is the parent ring system of the ladybird (or ladybug)
alkaloids (review: W.A. Ayer and L.M. Browne, Heterocycles,
1977, *7*, 685).
The parent perhydro-compound exists in two stereoisomeric

forms (23) and (24).

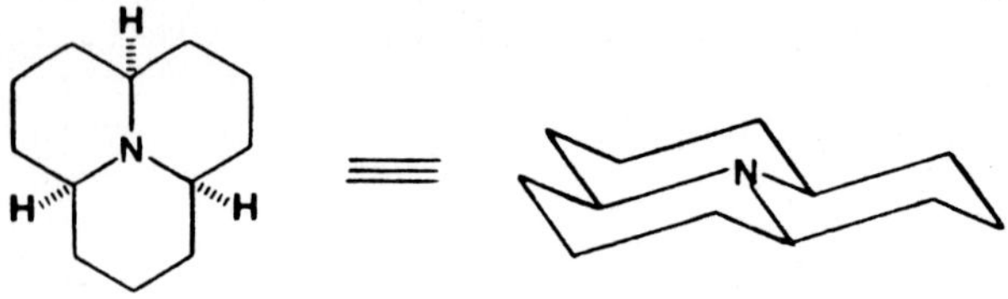

(23)

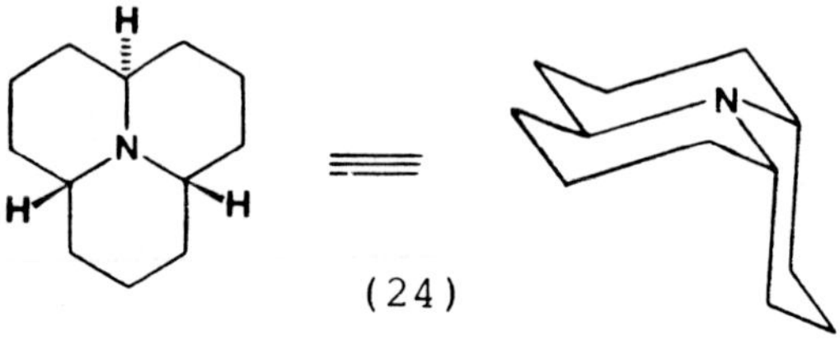

(24)

(3aα,6aβ,9aβ)-Dodecahydropyrido-[2,1,6-*de*]quinolizine (24) forms a picrate, m.p. 200-202°C, and is characterised by the presence of seven signals in its ^{13}C-nmr spectrum and by the absence of Bohlmann bands (2nd. edn. Vol IVH, p. 267) in its ir-spectrum. It is obtained from the perhydroboraphenalene

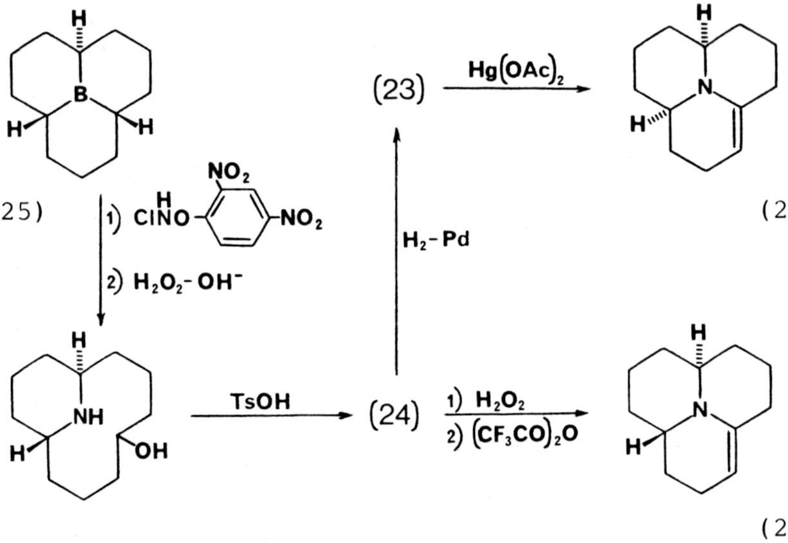

(25) by the route shown above and is converted into the more stable (3aα,6aα,9aα)-compound (23), picrate, m.p. 191-193°C, by exposure to hydrogen in the presence of a palladium catalyst. The latter isomer is characterised by the presence of only three signals in its ^{13}C-nmr spectrum and Bohlmann bands in its ir-spectrum (R.H. Mueller, M.E. Thompson, and R.M. DiPardo, J. org. Chem., 1984, 49, 2217).

The perhydro-compounds (23) and (24) may be converted into the enamines (26) and (27), respectively, and (26) is identical with the product obtained (I. Murakoshi et.al., Chem. pharm. Bull., 1964, 12, 747) by heating the quinolizidone (28) with soda-lime.

(28)   (29)   (30)   (31)

The enamines (26) and (27) are key intermediates in the synthesis of the ladybird alkaloids, as also are the stereo-isomeric ketones (29) and (30), prepared as a 1:1 mixture from the bis(ethylene acetal) of the disubstituted piperidine (31) (W.A. Ayer and K. Furuichi, Canad. J. Chem., 1976, 54, 1494). More importantly, the (3aβ,6aα,9aβ)-ketone (30), to which most of the alkaloids are stereochemically related, is obtained stereospecifically from the enamine (27) (R.H. Mueller et.al., loc.cit.) or by a particularly elegant route (shown below) based on the Robinson-Schöpf reaction (R.V. Stevens and A.W.M. Lee, J. Amer. chem. Soc., 1979, 101, 7032).

5. Bridged ring compounds

(a) Bicyclic systems having a nitrogen bridge

The chemistry of 7-azabicyclo[2.2.1]heptadienes is the subject of a review (L.J. Kricka and J.M. Vernon, Adv. heterocyclic Chem., 1974, *16*, 87). These compounds (1) are formed
in Diels-Alder reactions of pyrroles with acetylenic dienophiles, better yields being obtained when the pyrrole has an
electron-withdrawing *N*-substituent.

(1)   (2)   (3)

Rearrangements are important in the chemistry of azabicycloalkanes, as in that of their hydrocarbon counterparts.
An example of this is to be found in the formation (83%) of
the 8-azabicyclo[3.2.1]octene (tropidine) derivative (3) from
the isomeric [5.1.0]-compound (2) in the presence of bis-
(benzonitrile)palladium(II) dichloride (G.R. Wiger and M.F.
Rettig, J. Amer. chem. Soc., 1976, *98*, 4168). Rearrangements
are also observed when *N*-chlorotropane (4) and *N*-chloro-
granatanine (5) are treated with silver(I) fluoroborate in
benzene. After reduction of the intermediate iminium salts,
in situ, with sodium borohydride, these reactions yield (>90%)

(4)

(5)

pyrrolizidine and indolizidine, respectively (F.M. Schell and R.N. Ganguly, J. org. Chem., 1980, *45*, 4069).

In a formal violation of Bredt's rule, the 11-azabicyclo-[4.4.1]undecane ring system can exist in an aromatic form, 1,6-imino[10]annulene (6), a yellow solid, m.p. 16°C. This compound is prepared (75%) from the tricyclic aziridine (7) by addition of bromine (2 equiv.) and dehydrobromination. A ring current in the [10]annulene π-system of the compound causes deshielding of the CH proton signals (δ 7.2) in its nmr spectrum and strong shielding of the NH proton (δ -1.1) (E. Vogel, W. Pretzer, and W.A. Böll, Tetrahedron Letters, 1965, 3613).

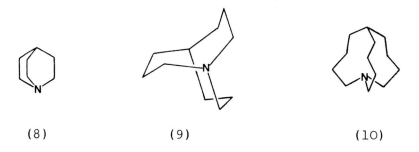

(6)   (7)

(b) *Bicyclic systems having nitrogen at a bridgehead position*

The 1-azabicyclo[n.n.n]alkanes form an interesting series in which to observe the effects of geometry on the donor properties of nitrogen. The first member of the series (n=1) is not yet known but the second member, 1-azabicyclo-[2.2.2]octane or quinuclidine (8) (2nd. edn. Vol. IVH, p. 281) is a normal tertiary amine (pK$_a$ 10.58) except in so far as the nitrogen lone pair of electrons is less sterically hindered (has stronger Lewis base properties) than that in an acyclic tertiary amine.

(8)   (9)   (10)

1-Azabicyclo[3.3.3]undecane (9), also known as manxine, m.p. 150-152°C, hydrochloride, m.p. 305-307°C, and 1-azabicyclo-[4.4.4]tetradecane (10), m.p. 168-171°C, are prepared by reduction of tne bridgehead-bonded quaternary salts (11) and (12), respectively, with sodium in liquid ammonia (J.C. Coll *et.al.*, J. Amer. chem. Soc., 1972, *94*, 7092; R.W. Alder and R.J. Arrowsmith, J. chem. Research, 1980, (S) 163, (M) 2301).

Br⁻

(11)

BF₄⁻

(12)

In order to minimise the strain energy due to transannular interactions between hydrogen atoms in the bridging chains, manxine adopts a conformation in which the bonding geometry around the nitrogen atom is almost planar.   Since protonation results in a partial restoration of pyramidal geometry (A.H-J. Wang *et.al.*, J. Amer. chem. Soc., 1972, *94*, 7092) with a consequent increase in strain energy, manxine (pK_a 9.9) is considerably weaker, as a base, than quinuclidine.   Other notable properties of manxine are the relatively high wavelength and intensity (240 nm, ε 2935) of its first electronic transition (J. Coll *et.al.*, *loc.cit.*; A.M. Halpern, J. Amer. chem. Soc., 1974, *96*, 7655) and low value (7.05 eV) of its vertical ionisation potential (D.H. Aue, H.M. Webb, and M.T. Bowers, *ibid.*, 1975, *97*, 4136).   Both of these features are atypical of tertiary amines and provide further evidence of flattening at the nitrogen atom.

In 1-azabicyclo[4.4.4]tetradecane (10), the bridging chains are sufficiently long to allow the nitrogen atom to return to its normal bonding geometry by becoming 'inwardly pyramidalised'.   The nitrogen lone-pair is thus relatively inaccessible within the cage-like structure and the compound is an exceedingly weak base (pK_a 0.6 in 48% aqueous ethanol; immeasurable in water) (R.W. Alder *et.al.*, *loc.cit.*; Chem. Comm., 1982, 940).

Chapter 38

THE QUINOLIZINE ALKALOIDS

MALCOLM SAINSBURY

Introduction

Quinolizine alkaloids are found in a wide variety of
leguminous plants and also, but to a lesser extent, in some
other unrelated plant families. They range in complexity
from simple bases such as lupinine (1) and the phenanthro
derivatives, illustrated by cryptopleurine (2), to tricyclic
and tetracyclic structures, of which cytisine (3) and
lupanine (4) respectively are well known representatives.
  There is evidence that in Nature tetracyclic alkaloids form
first and are then degraded to the tricyclic forms. Indeed
it is possible to discern the remnants of the fourth ring in
the structures of some tricyclic bases.
  Sometimes as in the case of camoensine (5) and its analogues
the fourth ring is five- rather than six-membered, and in the
Ormosia series the second nitrogen atom is no longer at a
bridgehead position. A typical alkaloid of this group is
ormosanine (6).
  A further class of quinolizine bases include the alkaloid
matrine (7). This compound and its stereomers are typically
found in plants of the *Sophora* genus, but the defence
secretions of beetles, notably ladybirds of the Coccinellidae,
also produce similar bases. Although strictly not true
alkaloids these azaphenalenes, e.g. coccinelline (8), are
probably "vegetable in origin" since their immediate
precursors are transmitted to the beetles *via* the food chains
of their prey.

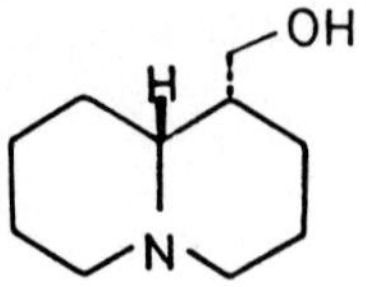
OH
H
N

1

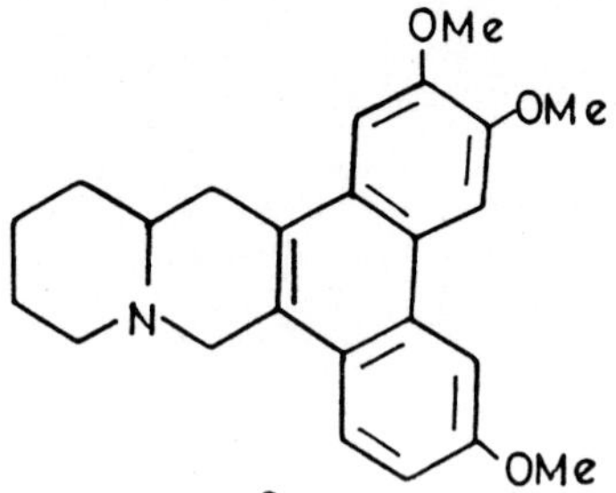
OMe
OMe
N
OMe

2

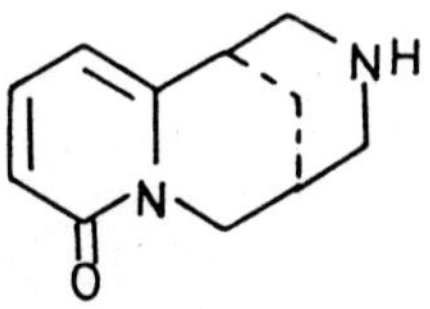
NH
N
O

3

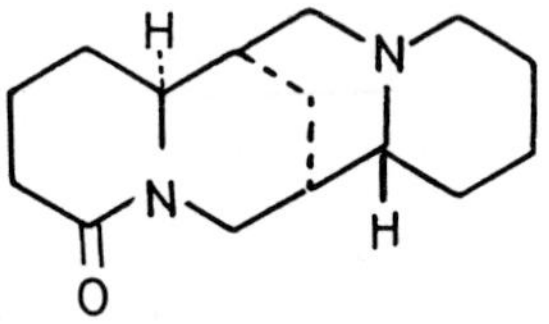
H
N
N
H
O

4

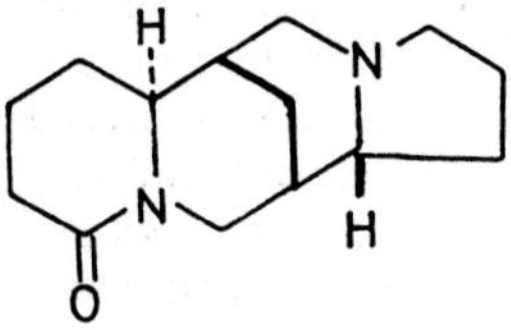
H
N
N
H
O

5

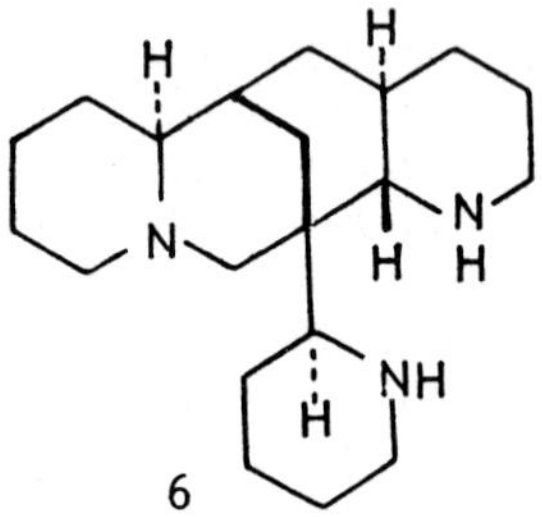
H
H
N
N
H
H
H
NH

6

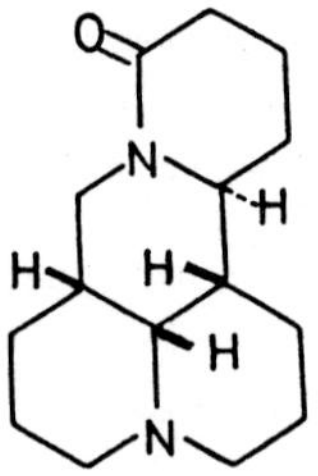
O
N
H
H
H
H
N

7

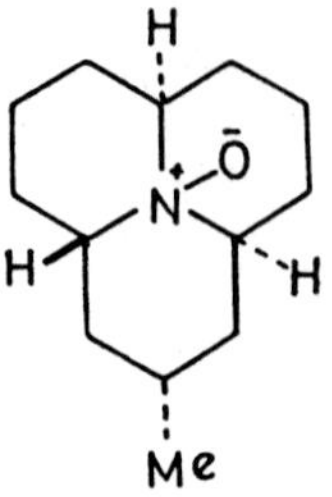
H
N
O
H
H
Me

8

1. *Lupinine alkaloids and related compounds*

(a) *New Structures*

Cis and *trans*-4-hydroxycinnamoyl esters (9, R=H) of
(-)-lupinine have been obtained from the seedlings of
Lupinus lutens. The *trans*-isomer is by far the more
abundant and often occurs with its 3-methoxylated derivative
(9, R=OMe) in the form of glucose and rhamnose glycosides
(I. Murakoshi *et al.*, Phytochem., 1975, *14*, 2714; 1977, 16,
2046; 1978, *17*, 1817; 1979, *18*, 699).

9

Epilamprolobine (10) occurs in extracts of *Sophora tomentosa*,
together with its *N*-oxide (11) and a ring-opened derivative
(12). Since, however, this last structure may be formed by
treating the *N*-oxide with methanol there is doubt that it is
a true natural product (W.J. Keller, *ibid.*, 1980, *19*, 2233;
Murakoshi, *ibid.*, 1981, *20*, 1725). Moreover, the same
suspicion must rest on the structure of sophorine (13)
claimed as an alkaloid of *S. alopercuroides* (F.G. Kamaev *et
al.*, Khim.prirod.Soedin., 1981, 604).

10

11

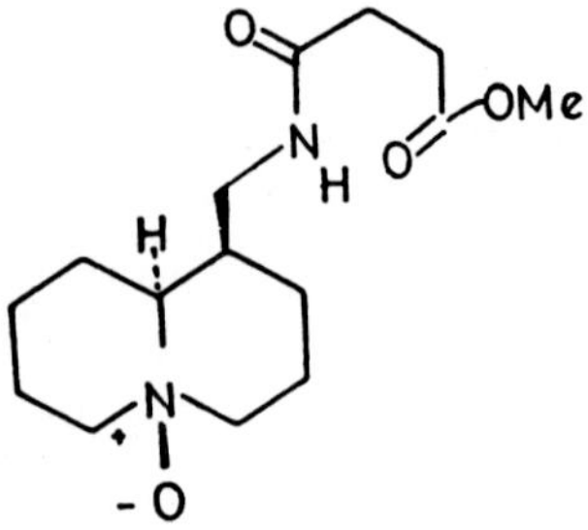

12

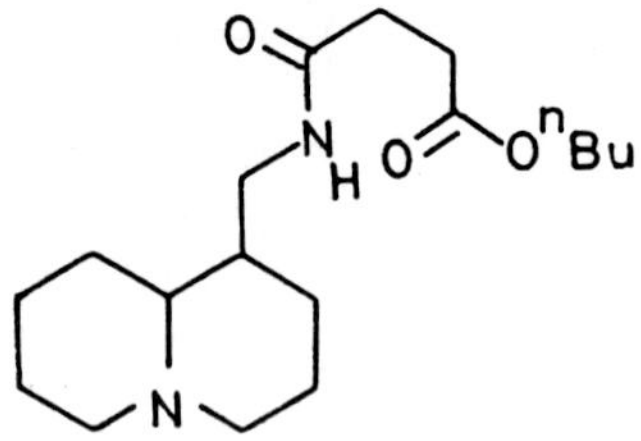

13

Lupinine bases are also found in plants of the Lythraceae family, an example is arbresoline (14, R=OMe) isolated from *Heimia salicifolia* some years ago. Desmethoxyarbresoline (14, R=H) and 10-epi-desmethoxyabresoline (15) are two new bases from the same source (A. Rother and A.E. Schwarting, Phytochem., 1978, *17*, 305). Four similar structures occur in *Lagerstroemia subcostata* (K. Fuji *et al.*, Chem.pharm.Bull., 1978, *26*, 2513). These are lasubine-I (16), lasubine-II (17), subcosine-I (18) and subcosine-II (19).

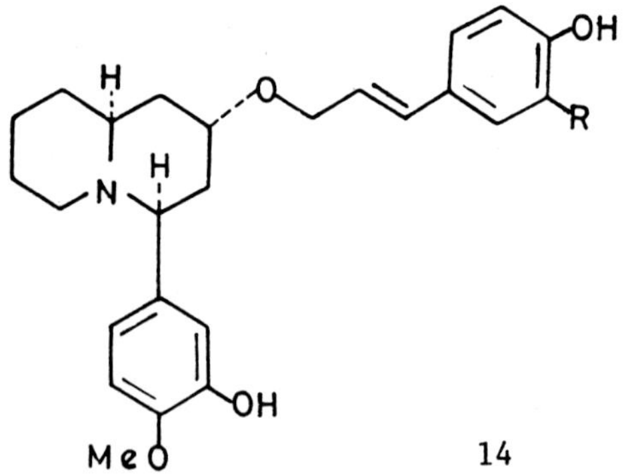

14

15

16

17

18

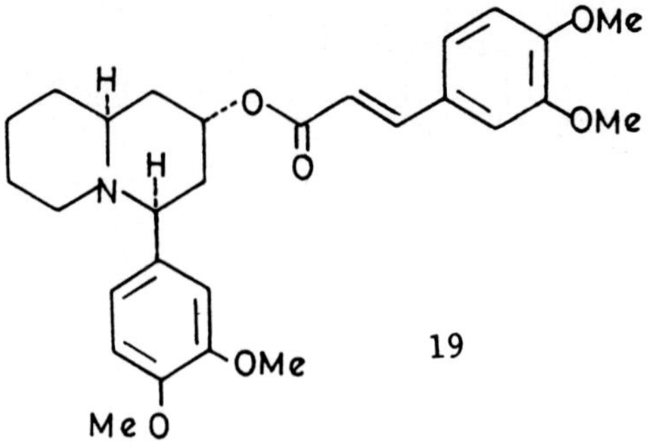

19

The water lily *Nuphar pumila* elaborates an unstable base
(+)-numpharopumiline which on catalytic reduction affords
the known alkaloid (-)-desoxynupharidine (21) from this
evidence, and from an analysis of the ^{1}H n.m.r. spectrum,
structure (20) is proposed for the new alkaloid (P. Peura
and M. Lounasmaa, Phytochem., 1977, *16*, 1122).

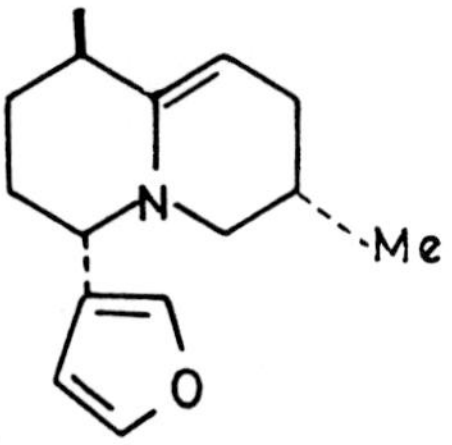

20

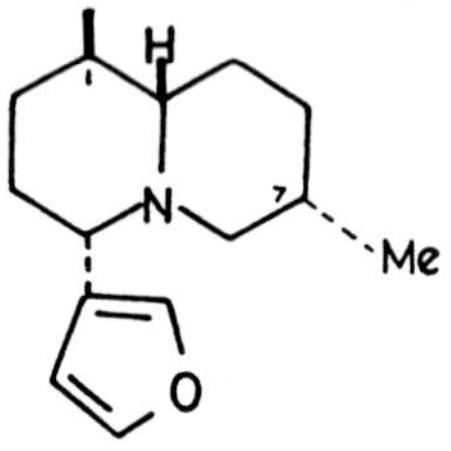

21

(-)-Desoxynupharidine and (-)-7-epidesoxynupharidine (22)
are known alkaloids from other *Nuphar* species, however,
it is interesting to note that they also co-occur with the new
bases   (-)-1-epidesoxynupharidine,   (-)-1-epi-7-epidesoxy-
nupharidine and (-)-isocastoramine (23) in the scent gland of
the Canadian beaver (*Castor fiber*) (B. Mauer and G. Ohloff,
Helv., 1976, *56*, 1169).

Petrosins-A (24) and -B (25) are stereoisomeric bisquinol-
izines from the sponge *Petrosia seriata* (J.C. Braekman *et al.*,
Bull.Soc.chim.Belg., 1984, *93*, 941). A third base from the
same source is petrosin (26). Petrosin-B, unlike the other two
compounds, lacks a two-fold axis of rotation.

24    $R^1 = R^4 = \alpha-H;$ $R^2 = R^3 = \beta-H$

25    $R^1 = R^3 = \alpha-H;$ $R^2 = R^4 = \beta-H$

26    $R^1 = R^4 = \beta-H;$ $R^2 = R^3 = \alpha-H$

(b)   Syntheses

A concise synthesis of (±)-lupinine is outlined in Scheme 1.
(G.C. Gerrans, A.S. Howard and B.S. Orlek, Tetrahedron
Letters, 1975, 4171). Starting from 2-thiopiperidine (27)
two alkylation sequences introduce *N*- and *C*-substituents
which ultimately form the second ring of the quinolizidine
system .

27

Scheme 1   Reagents: (a) $CH_2=CHCO_2Et/NaH$ (b) $BrCH_2CO_2Et$ /

Et_3N/Ph_3P (c) $LiAlH_3OEt$ / NaH/TsCl   (d) $NaBH_4$ /   LAH

(±)-Epilupinine (33) has been synthesised by the cyclisation
of the ketene thioacetal (28) by treatment with methane
sulphonyl chloride and triethylamine. It is presumed that
the acyliminium ion (29) is an intermediate in this reaction
and from it is formed the *S*-stabilized cation (30).
Deprotonation then affords the tricycle (31) which, after
deprotection in the presence of methanol gives the ester (32).
Conversion of this product into epilupinine is achieved by
reduction with lithium aluminium hydride (A.R. Chamberlin,
H.D. Nguyen and J.Y.L. Chung, J.org.Chem., 1984, *49*, 1682)
(Scheme 2).

Scheme 2 Reagents: (a) $MeSO_2Cl/Et_3N$ (b) Base
(c) $HgCl_2/MeOH$ (d) LAH

An alternative approach utilizing a silicon directed N-acyliminium ion cyclisation has been reported by H. Hiemstra et al (ibid., 1985, 50, 4014).

64

(±)-7-Epi-desoxynupharidine (37) and 1-epi-7-epidesoxy-
nupharidine have been prepared from 2-ethyl-5-methylpyridine
(34) *via* the ketone (35) and 3-formylfuran (36). The product
from this series of reaction is a mixture of four
stereoisomeric quinolizidines but, after treatment with
sodium methoxide in methanol, the more stable *trans* fused
compounds is isolated and reduced under Wolff-Kishner
conditions to the corresponding desoxy derivatives (S. Yasuda,
M. Hanaoka and Y. Arata, Heterocycles, 1977, *6*, 391).

Scheme 3 Reagents: (a) CH_3CN /PhLi (b) $HOCH_2CH_2OH/H^+$
(c) $H_2/Rh/H^+$ (d) MeONa (e) NH_2NH_2/B

(±)-Desoxynupharidine has also been synthesised (Y.C. Hwang and F.W. Fowler, J.org.Chem., 1985, *50*, 2719).
Myrtine (39) is a simple quinolizidinone from the heather *Vaccinium myrtillus,* its structure was deduced by spectroscopy and by a short synthesis from (-)-(*R*)-pelletierine (38) (see Scheme 4) (P.Slosse and C. Hootele, Tetrahedron Letters, 1978, 397; 1979, 4587).

Scheme 4   Reagents : (a) CH_3COCHO/pyridine (b) $Al(O^tBu)_3$
(c) MeMgI

Several known Lythraceae alkaloids including decaline, vert-aline (K. Shishido *et al.*, Chem.pharm.Bull., 1985, *33*, 532) and lythrancepine-II (D.J. Hart and W.P. Hong, J.org.Chem., 1985, *50*, 360) have been synthesised.

Cryptopleurine (42) shows anticancer activity and this has stimulated much interest in the synthesis of it and its analogues. One efficient route to both cryptopleurine and (±)-julandine (41) involves the common intermediate (40), prepared as shown in Scheme 5 . Reduction of this compound with lithium aluminium hydride leads directly to julandine, whereas oxidative cyclisation by ultraviolet irradiation in the presence of iodine followed by reduction of the amide group gives cryptopleurine (H. Iida and C. Kibayashi, Tetrahedron Letters, 1981, *22*, 1913; J.org.Chem., 1984, *49*, 2412). Equally interesting approaches to these alkaloids have been described by Herbert and his colleagues (*ibid.*, p.2127) and by Snieckus and co-workers (*ibid.*, p.2349).

40

41

41 $\xrightarrow{\text{f}}$

42

Scheme 5 Reagents: (a) Zn/HOAc (b) $4\text{-MeOC}_6\text{H}_4\text{CH}_2\text{COCl}/\text{K}_2\text{CO}_3$
(c) O_2 (d) NaOEt (e) LAH (f) $I_2/h\nu$

2. *Alkaloids of the cytisine type*

New structures

Some new alkaloids which illustrate the comment made in the
introduction concerning the biosynthetic interrelationship
between the three- and four-ring structures are the *N*-
butenylated base (43) from *Lupinus costentinii*,
N-(3-oxobutyl) cytisine (44) and *N*-ethylcytisine (45) from
Echinosophora koreensis (I. Murakosi *et al.*, Phytochem., 1982,
21, 1470), and the Zwitterion (46) extracted from *Euchresta
japonica* (S. Ohmiya *et al.*, *ibid.*, 1979, *18*, 649); see also
Murakosi *et al.*, Phytochem., 1985, *24*, 2707.
11-Allylcytisine (47) has been isolated from *Clathrotropsis
brachypetala* (G.M. Hatfield, W. Keller and J.M. Rankin,
J.nat.Prod., 1980, *43*, 164) and also from the unripe fruit
of *Sophora secundiflora* (Keller and Hatfield, Phytochem.,
1979, *18*, 2068) and a stereoisomer of the known alkaloid
tinctorine (48) has been obtained from *Baptisia australis*
(M. Wink *et al.*, J.nat.Prod., 1981, *44*, 14).

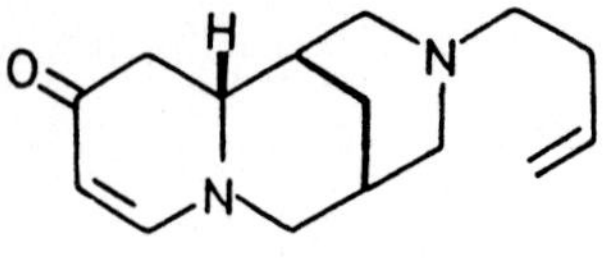

43

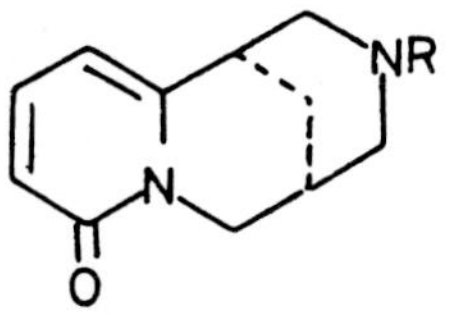

44,   R=CH$_2$CH$_2$COMe
45,   R=Et

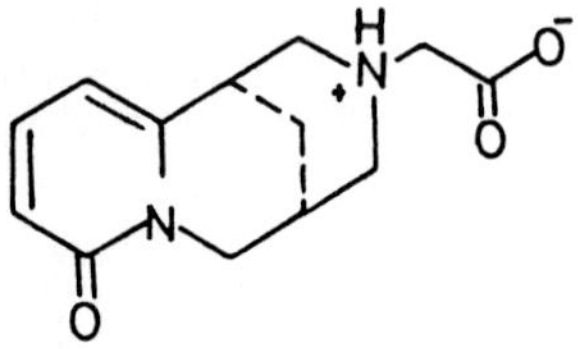

46

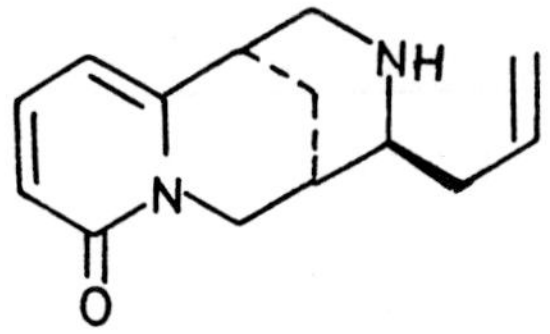

47

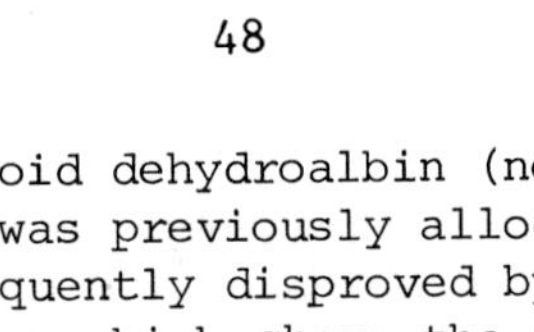

48

The tricyclic alkaloid dehydroalbin (now better named albine)
from *Lupinus albus* was previously allocated structure (49).
This has been subsequently disproved by an X-ray crystall-
ography determination which shows the alkaloid to be the
reduced form (50) (A.N. Chekhlov *et al.*, Zhur.strukt.Khim.,
1974, *15*, 950; A. Hoser *et al.*, Acta Crystallogr., 1980, *36*,
984). Virgiboidine isolated from *Virgillia oroboides* and
V. divaricata contains an *N*-butenyl substituent, but the
position of the carbonyl group located in structure (51) at
C-10 is not totally secure (J.L. van Eijk and M.H. Radema,
Planta Med., 1982, *44*, 224).

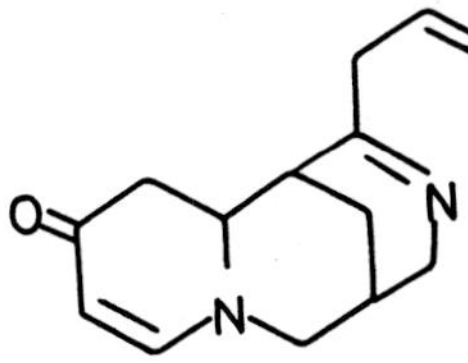

49

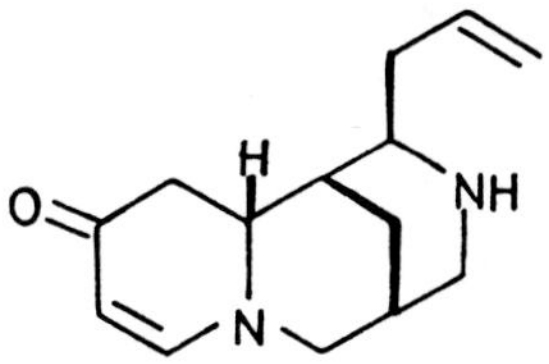

50

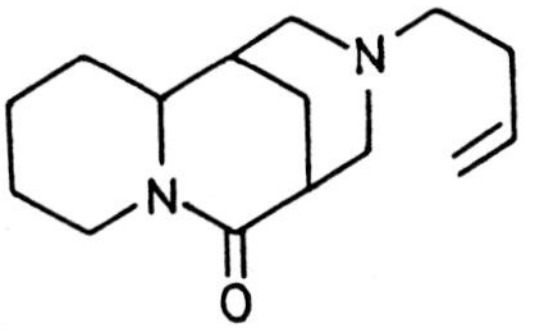

51

The tsukushinamines-A (52), -B (53) and -C (54), extracted
from *Sophora franchetiana* (S. Ohmiya *et al.*, Chem.pharm.Bull.,
1979, *27*, 1055; 1980, *28*, 1965; Phytochem., 1981, *20*, 1997),
are included here for it seems likely that they are
biosynthesised by cyclisation to C-6 of the *N*-alkyl
substituent of a cytsisine-like precursor.

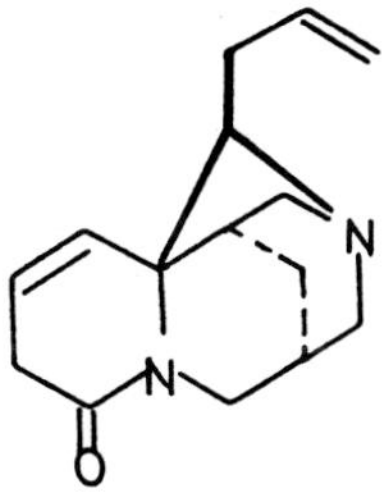

52

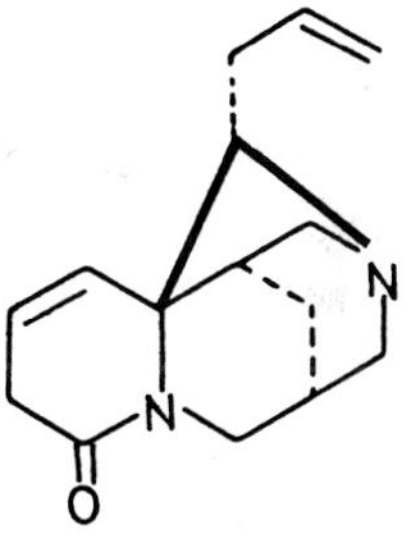

53

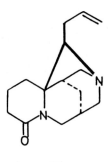

54

3. *Lupanine alkaloids*

New compounds

Esters of 13-hydroxylupanine (55) are common among the extractives of the *Genista, Lupinus* and *Sarothaminus* (syn. *Cytisus*) genera Some of these compounds are listed in the main work p. 308.

Cadiamine (56) and the esters (57) and (58) occur together in *Cadia purpurea* (J.L. van Eijk, M.H. Radema and C. Versluis, Tetrahedron Letters, 1976, 2053). Such structures probably arise in Nature through fragmentation of a preformed tetracyclic lupanine base. (-)-Mamanine-*N*-oxide (59) isolated from *Sophora chrysophylla,* together with eighteen other alkaloids of known structure, may have a similar origin (I. Murakoshi *et al.,* Phytochem., 1984, *23*, 887).

55

56, R=H
57, R=2-pyrroloyl
58, R=4-hydroxyphenylacetyl

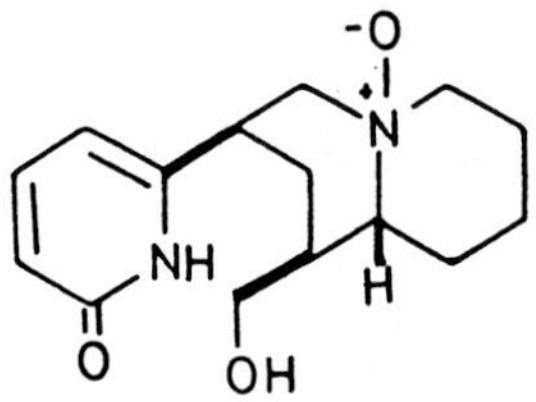

59

"Isotinctorine" (referred to in section 2) is accompanied in
Baptisa australis by 13-acetoxyanagyrine (60) or a stereoisomer of this structure (M. Wink *et al.*, J.nat.Prod., 1981,
44, 14).   A hydroxy derivative of anagyrine (structure
uncertain) occurs in *Anagyris foedida* (J.M.V.-Lobo *et al.*,
An.Quim., 1977, *73*, 1366).

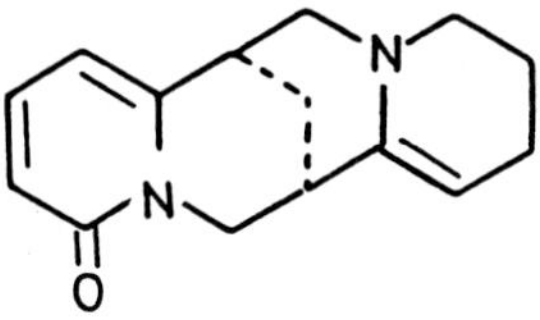

60

Chamaetine (61) from *Chamaecytisus* species is a stereoisomeride of the known alkaloid nuttaline (62) (A. Daily *et
al.*, Tetrahedron Letters, 1978, 1453).

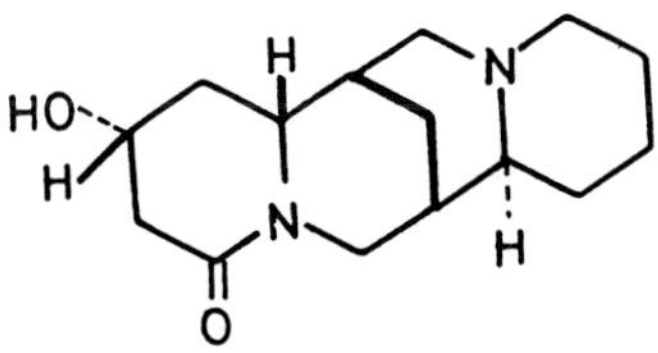

61

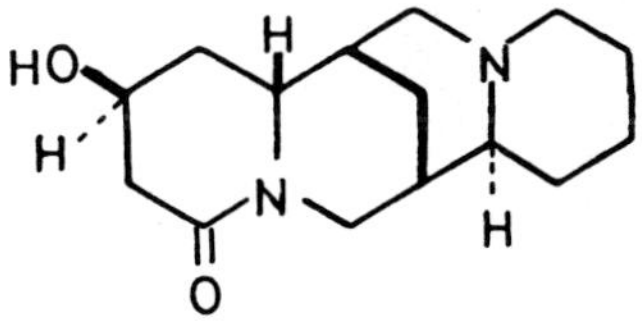

62

4,13-Dihydroxylupanine (63) is present in the plant
Sarothamnus scoparius together with a structure considered
to be 11,12-dehydrosparteine (64) (Wink, L. Witte and T.
Hartmann, Planta Med., 1981, *43*, 342). An isomer of the
first alkaloid, 10,13-dihydroxylupanine (65) accompanies
13-hydroxylupanine in *Cadia purpurea* (J.L. van Eijk, M.H.
Radema and C. Versluis, Tetrahedron Letters, 1976, 2053).
Another alkaloid of this plant is said to be 13-
ethoxylupanine (Radema, Planta Med., 1975, *28,* 143), but
ethoxy groups in natural products are very unusual and
hence the true alkaloid may be 13-hydroxylupanine.

63

64

65

4.   *Quinolizidine - indolizidine alkaloids*

New Structures

Camoensine (66, R=H) and its reduction product camoensidine
(67, R=H) are established alkaloids of *Camonesia maxima*.
They also occur in *C. brevicalyx* together with three other
bases, two of which are 12α-hydroxycamoensine (66, R=OH) and
12-hydroxycamoensidine (67, R=OH). The third natural
product may have structure (68) (P.G. Waterman and D.F.
Faulkner, Phytochem., 1982, *21*, 215).

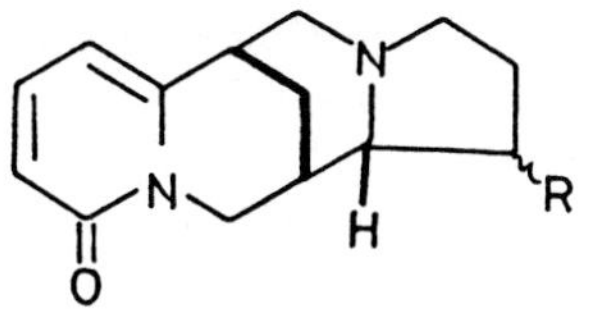

66

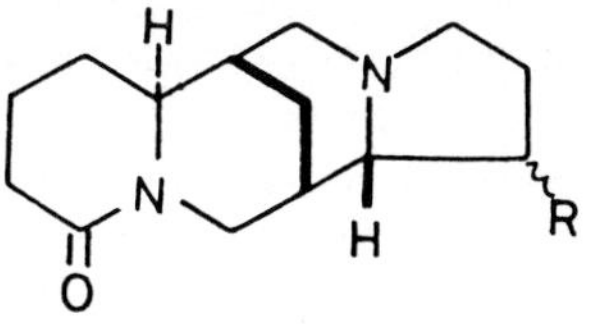

67

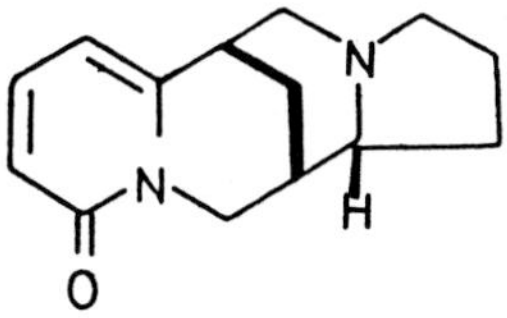

68

The structure of 11-epi-leotidane (69) a new alkaloid from
Maackia amurensis is confirmed since it is formed by reduction
of camoensidine with lithium aluminium hydride (A.D. Kingshorn,
M.F. Balandrin and L.-J. Lin, *ibid.*, 1982,*21*, 2269).

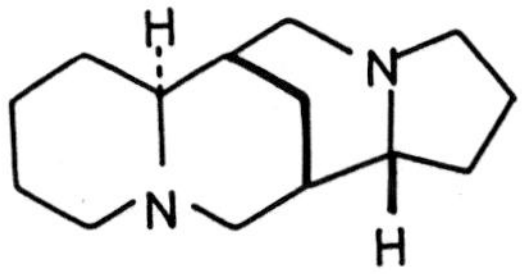

69

5. *Ormosia alkaloids*

New Structures

Since the publication of the second edition less than a
handful of new *Ormosia* alkaloids have been discovered, but it
has been noted that certain known bases tend to crystallise
together in the form of quasi racemates. Thus (-)-podopetaline
(70) and (-)-ormosanine (71) separate out together from

74

extracts of *Podopetalum ormondii* (R. Misra *et al.*, Chem.Comm., 1980, 659; S. McLean *et al.*, Canad.J.Chem., 1981, *59*, 34). 6-Epi-podopetaline (72), in the form of a hydrated hydrobromide salt, also occurs in this plant (M.F. Mackay and B.J. Poppleton, Cryst.Struct.Comm., 1980, *9*, 805).

An optically active form of jasmine (73) plus its stereoisomeride homodasycarpine (74) have been extracted from *Ormosia costulata* (J.K. Frank *et al.*, Acta Crystallogr., 1978, *34B*, 2316; A.H.J. Wang *et al.*, *ibid.*, p.2319).

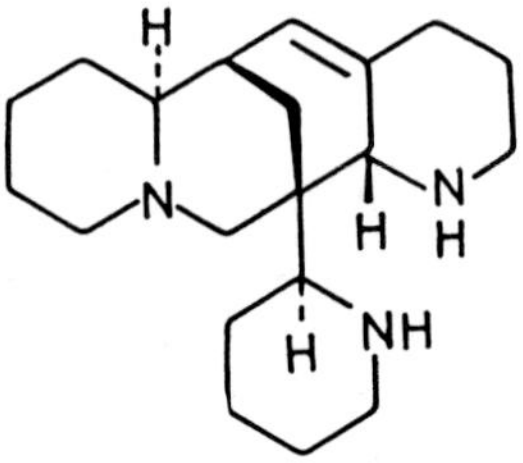

70

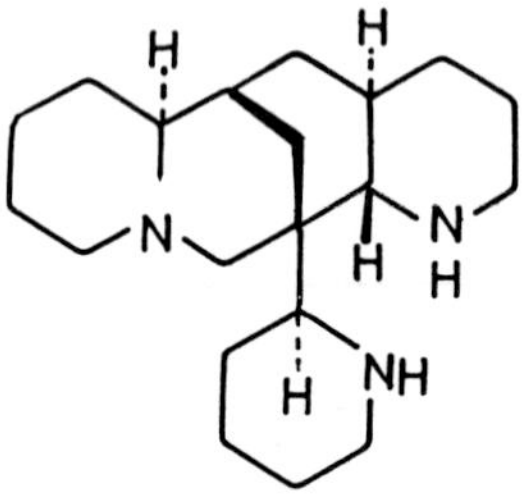

71

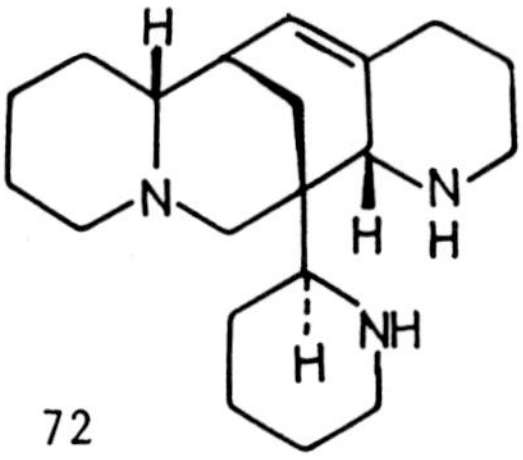

72

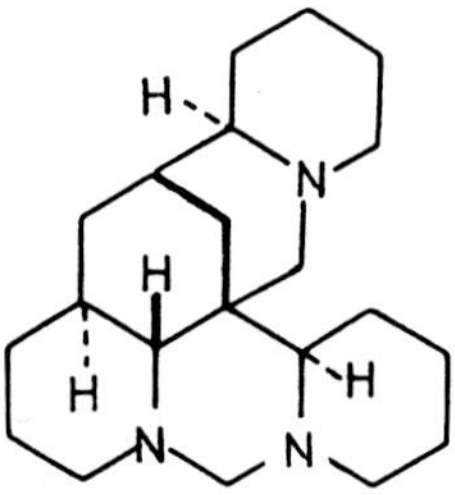

73

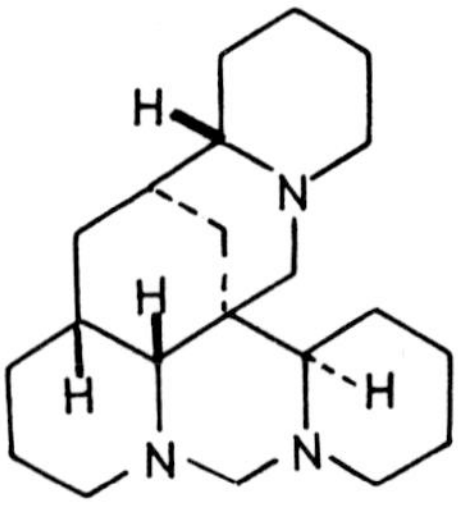

74

Aloperine has been allocated structure (75, R=H) some forty years after its isolation from *Sophora alopercuroides*. The same source yields the allyl derivative (75, R=CH$_2$CH=CH$_2$) (O.N. Tolkacher *et al.*, *ibid.*, 1975, 30).

75

Sweetinine (76) occurs in both *Sweetia panamensis* and *S. elegans*. This structure has some similarities to those of nitraramine (77, R=H) and N-hydroxynitraramine (77, R=OH) from *Nitraria schoberi* (N. Yu Novgorodova, S. Kh. Maeckh and S. Yu Yunusov, Khim.prirod.Soedin., 1975, 529).

76

77

6. *Matrine-like alkaloids and their congeners*

In addition to the lupinane group of alkaloids lupin plants and their allies produce a series of isomeric bases which take their name from matrine, one of the first of these alkaloids to be characterised. The relationship between the two groups is close and they share the same biosynthetic origin - thus they probably reflect alternative ring-closures of a hypothetical common intermediate (78).

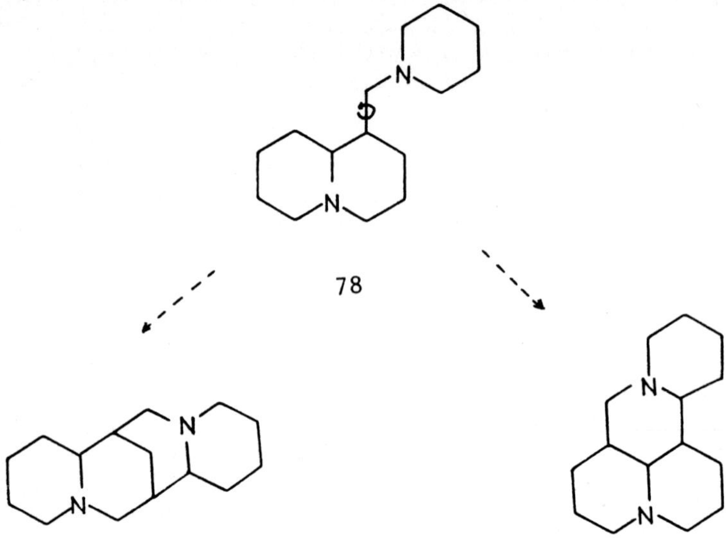

78

Lupanine series   Matrine series

New structures

Leontismine (79) from *Leontice smirnowii* has not previously
been described (E.G. Tkeshelashvili *et al.*, C.A., 1973, *78*,
156 646). Its structure is proven since when dehydrated the
new base affords isosophoramine (80), and when it is reduced
with lithium aluminium hydride leontane (81) is obtained.

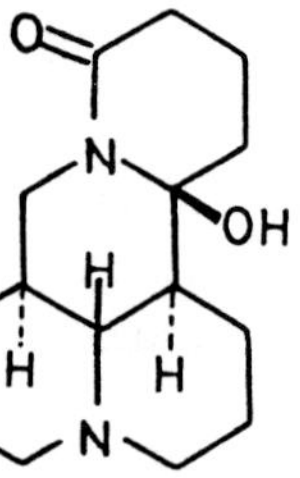

79 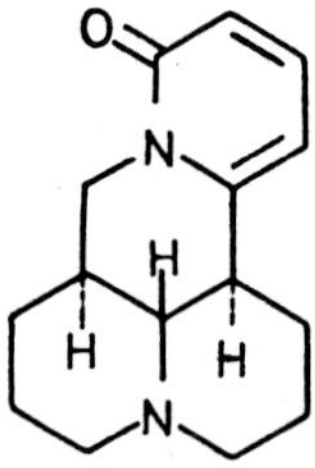   80 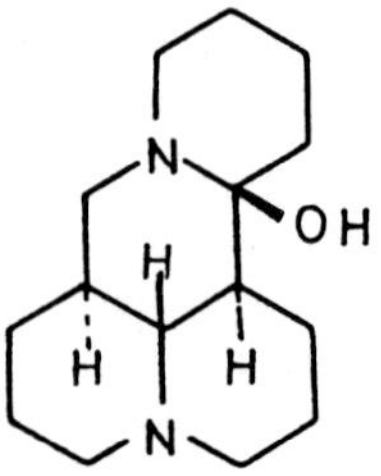   81

Hovea species not only give rise to tetracyclic alkaloids of the lupanine and sparteine type but also yield Ormosia type structures. For example, *H. Linearis* elaborates 6-epi-podo-petaline, (±)-piptanthine (82) and a third base which may be a stereoisomer of (-)-16-epi-ormosanine (83) (J.A. Lamberton, T.C. Morton and H. Suares, Austral.J.Chem., 1982, *35*, 2577).

Mass spectrometric comparisons indicate that neosophoramine from *Sophora* species is a stereoisomer of sophoramine (84), and from infra red and n.m.r. spectral evidence it is likely that it has structure (85) (T.E. Monakhova *et al.*, Khim. prirod.Soedin, 1974, 472). It has been established that (-)-sophoridine has the stereochemistry shown in formula (86) (F.G. Kamaev *et al.*, *ibid.*, p.7444), thus the structure of its 3-hydroxy derivative, previously isolated from *S. aloecuroides* (Monakhova *et al.*, *ibid.*, 1973, *9*, 59), must be revised.

Interestingly *N*-oxidation of (-)-sophorodine yields a mixture of *N*-oxides one of which is the minor alkaloid (87) of *Euchresta japonica* (S. Ohmiya *et al.*, Chem.pharm.Bull., 1980, *28*, 546).

S. flavescens is a source of three new alkaloids: 13,14-dehydrosophorodine (88) (Yu.K. Kushmuradov, S. Kuchkarov and Kh.A. Aslanov, Khim.prirod.Soedin., 1978, 231; A. Veno *et al.*, Chem.pharm.Bull., 1978, *26*, 1832), 7,8-dehydrosophoramine (89) (K. Morinaga *et al.*, *ibid.*, p.2483), (-)-7,11-dehydromatrine (90) (I. Murakoshi *et al.*, Phytochem., 1982, *21*, 2379).

5α,9 α-Dihydroxymatrine (91) is one of several alkaloids from *Eurchresta horsfeldii* (S. Ohmiya *et al.*, Phytochem., 1979, *18*, 645). Its structure was deduced from spectral data.

82

83

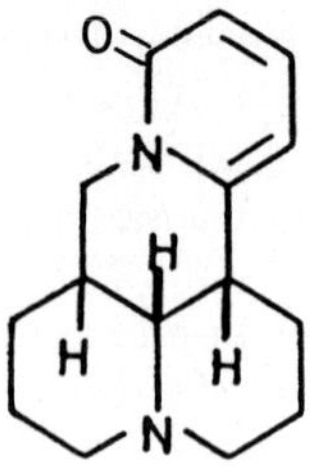

84

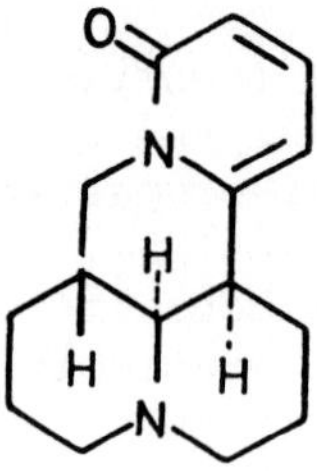

85

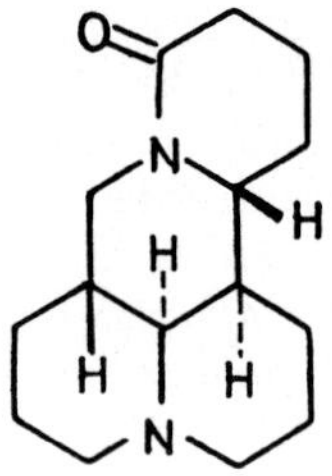

86

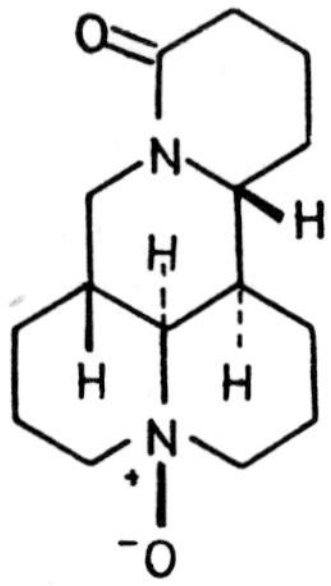

87

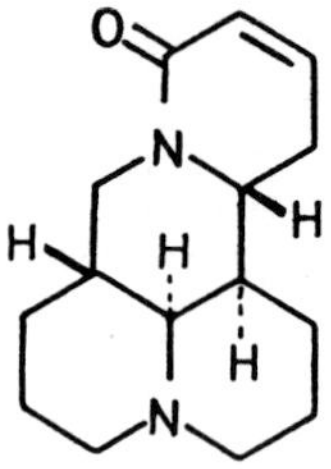

88

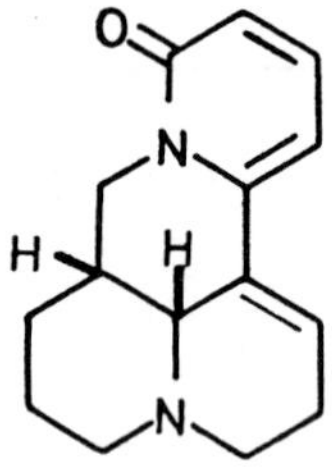

89

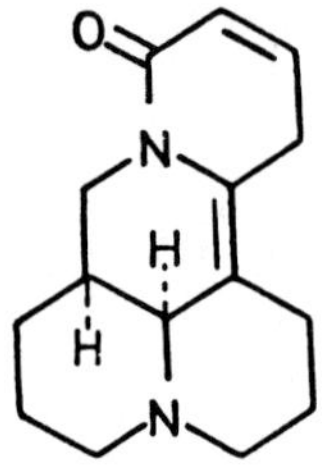

90

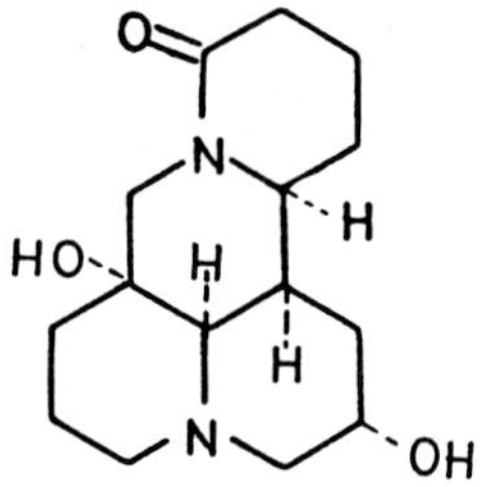

91

7. *9b-Azaphenalene alkaloids*

(a) *The bases*

Ladybird beetles of the family Coccinellidae secrete several
perhydro-9*b*-azaphenalene bases including coccinelline (92),
precoccinelline (93), hippodamine (94) and its *N*-oxide
convergine (95) (B. Tursch *et al.*, Experientia, 1971, *27*,
1380; Bull.Soc.chim.Belges, 1972, *81*, 649; Tetrahedron
Letters, 1973, 201; *ibid.*, 1974, 409; R.D. Henson *et al.*,
Experientia, 1975, *31*, 145; W.A. Ayer *et al.*, 1976, *54*, 1807).

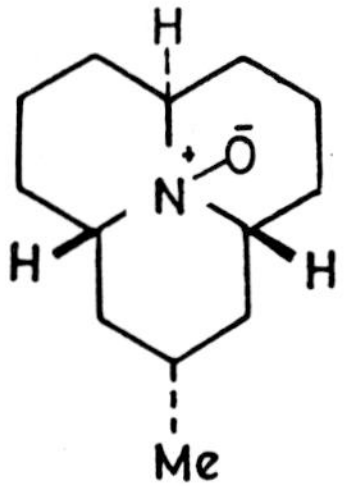

92

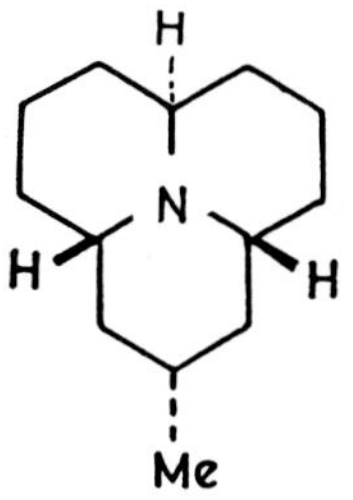

93

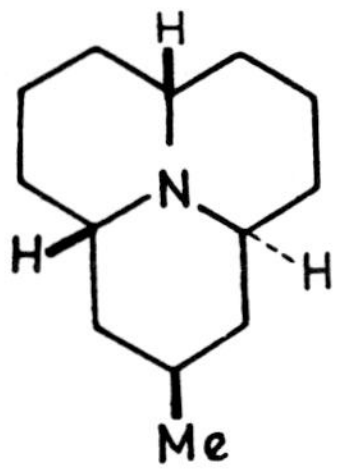

94

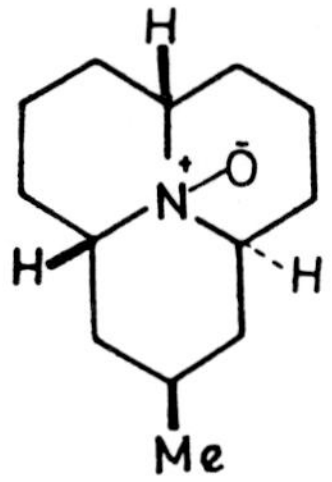

95

Myrrhine (96) from *Myrra octodecimguttata* is the all *trans*
isomer   of precoccinelline (Tursch *et al.*, Tetrahedron,
1975, *31*, 1541).   Indeed if coccinelline is treated with
ethyl chloroformate and the product (97) is then reduced a
mixture of precoccinelline and myrrhine is formed.

96

97

Hippocasine (98) and its *N*-oxide are decahydro-9b-azaphen-
alenes from *Hippodamia caseyi* (Ayer *et al.*, *loc.cit.*),while
propylein (99) is an unstable base from *Propylaea quatuor-
decimpunctata*. Hydrogenation of propylein affords

precoccinellin, but since its ultraviolet spectrum shows only
end absorption the lone pair of electrons on the N-atom must
lie out of conjugation with the Π-system of the double bond.
Thus propylein must have the conformation (100) (Tursch, D.
Daloze and C. Hootels, Chimia, 1972, *26*, 74).

98

99

100

Six related structures, porantherine (101), porantheridine
(102), poranthericine (103), *O*-acetylporanthericine (104),
porantheriline (I05) and porantherilidine (106) occur in
Poranthera corymbosa (S.R. Johns *et al.*, Austral.J.Chem.,
1975, 27, 2025; E. Gossinger, Tetrahedron Letters, 1980, *21*,
2229; Monatsh., 1980, *111*, 143).

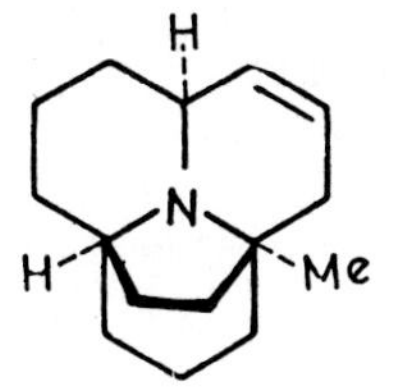

101

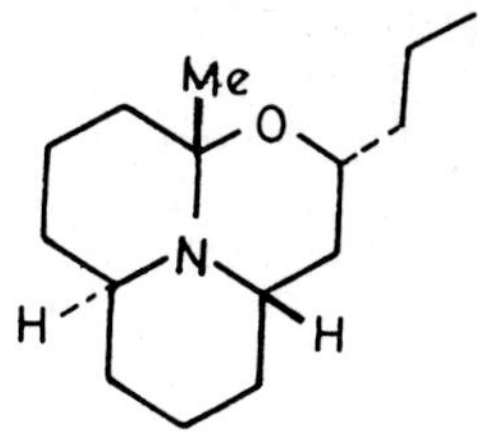

102

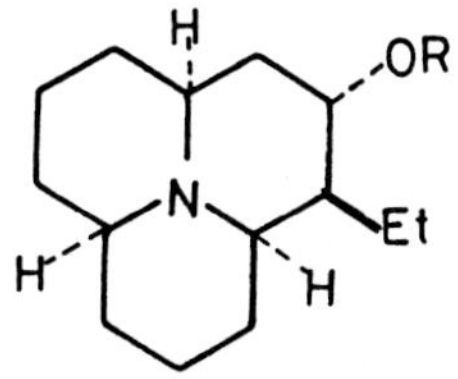

103   R = H
104   R = Ac

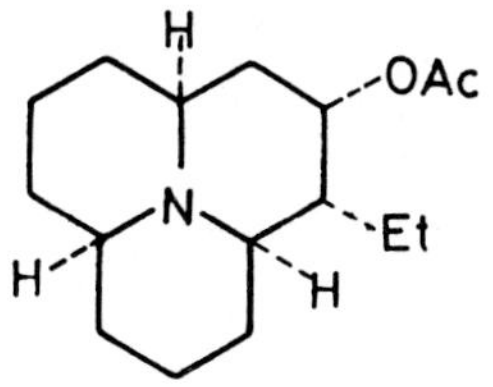

105

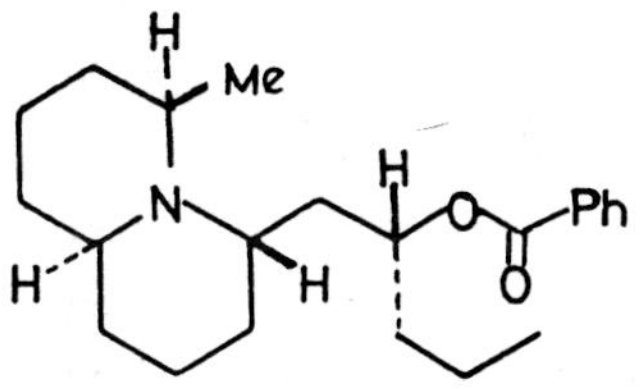

106

(b) *Syntheses*

A synthesis of porantherine, the only bridged structure of
this group, has been announced by E.J. Corey and R.D. Balanson
(J.Amer.chem.Soc., 1974, 96, 6516). In this work the 2-
piperideine (107) was reacted with isopropenyl acetate to

afford the ketone (108) which in two steps was converted into the aldehyde (109). This substrate was then cyclised to the enamine (110) and thence by heating with P-toluene sulphonic acid to the tetracyclic ketone (111). The formation of the alkaloid was then completed by a reduction-dehydration sequence (Scheme 6).

By any criterion this is an elegant synthesis, made the more notable because it represents an early example of a computer aided approach.

107

108

109

110

111

Porantherine

Scheme 6 Reagents: (a) isopropenyl acetate/H^+ (b) CrO_3/pyridine (c) OsO_4 (d) $HOCH_2CH_2OH/H^+$ (e) OH^- (f) HCl (g) H^+ (h) $NaBH_4$ (i) $SOCl_2$

Porantheridine and porantherilidine have been synthesised from
the same starting compound (112) (Scheme 7) (E. Gossinger,
Monatsh., 1980, *111*, 783). Selective oxidation of the secondary
alcohol group, protection by acetalisation and ring opening of
the isoxazolidine system, followed by treatment with benzoic
acid, triphenylphosphine and diethyl azodicarboxylate afforded
the benzoate (113). This substrate on debenzoylation and
deacetalisation yielded porantheridine, whereas when it was
simply treated with acid and then reduced porantherilidine and
its C-6- epimer were obtained.

112

113

e,f

Porantheridine

Porantherilidine

Scheme 7 Reagents : (a) CrO_3/H_2SO_4 (b) $HOCH_2CH_2OH/H^+$
(c) H_2/Ni (d) $EtO_2CN\ NCO_2Et/Ph_3P/PhCO_2H$ (e) KOH/MeOH
(f) H^+ (g) $NaBH_3CN$

Myrrine and hippodamine have been synthesised by W.A. Ayer
et al., (Canad.J.Chem., 1976, *54*, 473), while the latter
alkaloid, together with hippocasine and propyleine have been
prepared from the decahydro-9b-azaphenaline (114). This

compound was converted in five steps into the ketone (115) which, after reduction and dehydration yielded hippocasine. The same ketone, after thioacetalisation and reductive desulphurisation, gave hippodamine

The third base propylein was also obtained from the ketone by reduction and O-mesylation, followed by base promoted elimination of methanesulphonic acid. Isopropyleine was also produced in this sequence of reactions, a fact which indicates the iminium cation (116) to be a common intermediate (R.H. Mueller and M.E. Thompson, Tetrahedron Letters, 1980, 21, 1093; 1097).

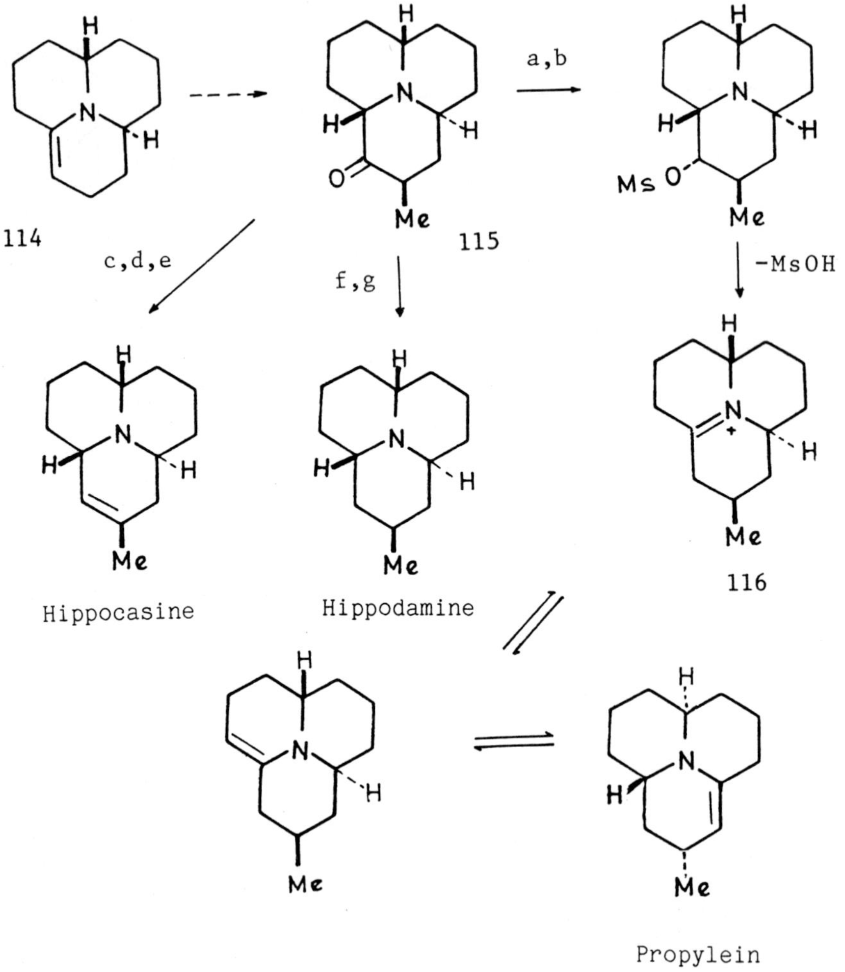

114

115

Ms OH

Hippocasine

Hippodamine

116

Propylein

Scheme 8 Reagents: (a) LAH (b) MsCl/Et$_3$N (c) NH$_2$NH$_2$
(d) p-TsCl (e) LiNH^{t}Bu (f) HSCH$_2$CH$_2$SH/BF$_3$ (g) Li/EDA

This approach has been developed further to allow stereo and regioselective total syntheses of (±)-precoccinelline, (±)-coccinelline, (±)-convergine and the *N*-oxide of hippocasine (Mueller *et al.*, J.org.Chem., 1984, *49*, 2217).

Chapter 39

COMPOUNDS CONTAINING TWO FUSED FIVE- AND SIX- MEMBERED
HETEROCYCLIC RINGS EACH WITH ONE HETERO ATOM

DEREK T. HURST

1. Compounds containing two hetero rings fused to an aromatic system.

(a) *Furoquinolines*

A versatile annelation route for the synthesis of furo-
[2,3-g]- and -[3,2-g]quinolines uses the reaction of regio-
isomeric bifunctional pyridine derivatives with vicinal
bromomethyl and (phenylsulphonyl)methyl groups to give substituted quinolines followed by their further reaction.
     Furo[2,3-g]quinoline (4), m.p. 115°, 70%,and furo[3,2-g]-
quinoline (3), m.p. 85°, 76%, are obtained in a one-operation
acid-induced cyclisation-elimination sequence from the
substituted quinolines (1) and (2) respectively (E. Ghera,
Y.Ben-David and H.Rapoport, J.org.Chem., 1983, 48, 774).

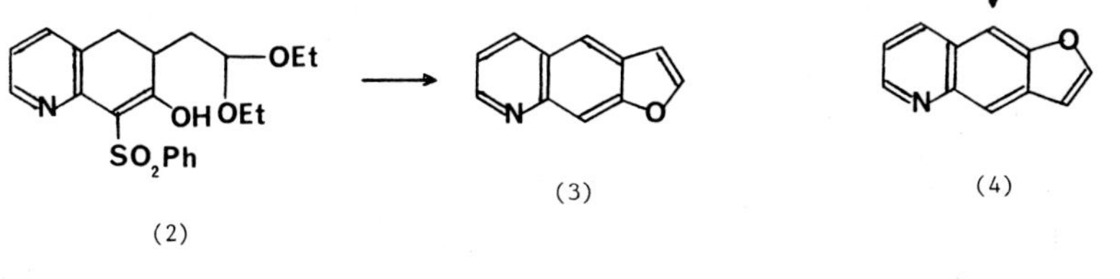

(b) *Pyrroloquinolines*

Methoxatin (10), a cofactor for methanol dehydrogenase in methylotrophic bacteria, has been synthesised. Key steps include the Fremy's salt oxidation of (5) to give the quinone imine (6) then condensation to give (7). Conversion of (7) into the chloroquinone (8) and subsequent trifluoro-acetolysis gives the oxazolinone (9) which is then converted into (10) (G. Büchi *et al.*, J.Amer.chem.Soc., 1985,107,5555).

(5)   (6)   (7)   (8)

(10)   (9)

(c) *Phenanthrolines*

N-Methyl-3,8-phenanthroline-5,6-dicarboxamide (13), m.p. 283°, 42% (isolated as the monohydrate) is obtained by a light-initiated dehydrobrominative cyclisation of dibromo-*N*-methylmaleimide (12) with 4,4'-bipyridyl dihydrate (11) (K.M. Wald *et al.*, Ber., 1980, 113, 2884).

(11)   (12)   (13)

The [3.3]cyclophane (14), m.p. 200°, containing the 1,10-phenanthroline system has been synthesised via a sequence of reactions involving ∝-functionalisation of 2,9-dimethyl-1,10-

phenanthroline (G.R. Newkome *et al.*, J.org.Chem.,1983, <u>48</u>, 5112).

(14)

A pentaethylene glycol crown ether derivative of 1,10-phenan-throline has also been synthesised (Newkome *et al.*, *ibid.* The macrocycles (15) and (15′) containing the 1,10-phenanthro-line-2,9-diphenyl moeity have been synthesised (C.O.Dietrich-buchecker and J.P. Sauvage, Tetrahedron Letters, 1983, 5091).

(15) n = 1
(15′) n = 2

A general method for the synthesis of dibenzo[b,j][x,z]-phenanthrolines where [x,z] is 1,7; 4,7 or 1,10 is exempli-fied by the reaction of 2-aminoacetophenone or 2-aminobenzo-phenone with 1,3-, 1,2-, or 1,4-diiodobenzene. The dibenzo-phenanthrolines (16), (17), (18) and (19) with fusion patterns [b,j][1,7], [b,j][1,7], [b,j][1,10] and [b,j][4,7] respectively are obtained either directly or after acid catalysed cyclisation of the diphenylamine intermediates (D. Hellwinkel and P. Ittemann, Ann., 1985, 1501).

(16) R = Me, m.p. 212–212.5°
(17) R – Ph, m.p. 268-268.5°

(18) m.p. 407–409°

(19) m.p. 211.5–212°

2. Compounds containing two hetero rings fused through adjacent carbon atoms.

(a) *Furopyrroles*

Aryl substituted furo[3,2-b]pyrroles are obtained by the thermolysis of azidofurylacrylic esters (made from 5–aryl-2-furylaldehydes and ethyl azidoacetate). This is an easy route to such compounds in high yield as shown for the compounds (20) below (A.Krutosikova, J.Kovac and J.Kristof-cak, Coll. Czech. chem.Comm., 1979, 44, 1799).

(20) R = Ph,   m.p. 169°
2-$NO_2C_6H_4$, m.p. 165°
3-$NO_2C_6H_4$, m.p. 210°

This reaction has been extended to give other substituted furopyrroles including *N*-alkylated compounds (A.Krutsikova *et al.*,Coll. Czech. chem.Comm., 1979, 44, 1808; 1980, 45, 2949; 1981, 46, 2564; 1983, 48, 772) and also to give benzofuro[3,2-b]pyrrole derivatives (A.Krutsikova *et al.*, *ibid.*, 1982, 47, 3288).

(b) *Furopyridines*

Furo[2,3-c]pyridine has been synthesised from 3-furoic acid chloride via the aldehyde and β-(3-furyl)acrylic acid azide (21) which, when heated at 180° in benzylamine gives the furo-pyridone (22). This compound is converted into the corresponding chloro compound by phosphorus oxychloride and on reduction with zinc/acetic acid gives the parent hetero-cycle, b.p. 150° at 40mm, methiodide m.p. 165-166° (S.Shiotani and H.Morita, J.heterocyclic Chem., 1982, 19, 1207).

(21)   (22)

A one-step synthesis of a mixture of furo[3,2-c]-(26) and furo[3,2-b]pyridines (25) is achieved by cycloaddition of an alkynic compound with 3,5-dichloropyridine 1-oxide. The formation of (25) probably proceeds through the 1,2-dihydro-pyridine (23) with a subsequent 1,5-sigmatropic shift to give (24) from which elimination of HCl gives (25). The formation of (26) is more difficult to understand (R.A.Abramovitch and I.Shinkai, J.Amer.chem.Soc., 1975, 97, 3227).

(23)   (24)

(26)   (25)

Either O,2- or O,4-cycloadditions of arylbromo- or -chloro-ketenes to 3-oxidopyridinium betaines give 2-oxofuro[2,3-c]-or 2-oxofuro[3,2-b]pyridines (A.Katritsky et al., Z.Chem., 1979, 19, 20).

The elaboration of the furan ring in a Feist–Benary reaction
starting from 4-hydroxy-6-methylpyran-2-one (27) is useful
for the synthesis of furo[3,2-c]pyridines (28) (E.Bisagin,
A.Civier and J.-P.Marquet, J.heterocyclic Chem., 1975, 12,
461. c.f. E. Bisagin *et al*., Bull.Soc.chim.France, 1971,
4041).
Furo[3,2-c]pyridin –3-ols (29) may be made in a similar
reaction (G. Hörlein *et al*., Ann., 1979, 371).

Other routes to the furopyridine ring system include cyclo-
dehydration of (30) to give (31) (W.E. Hymans and P.A.
Cruickshank, J. heterocyclic Chem., 1974, 11, 231) and the
Friedlander reaction of 3-amino-2-formylfuran (Scheme 1) (S.
Gronowitz and B. Uppström, Acta chem. Scand., 1975, 29B,224.
S. Gronowitz and A. Maetesson, *ibid*., 1975, 29B, 233).

(30)   (31)

(a)(i),NaN₃,DMSO (ii),H₂S,MeOH

(b)(i),MeCOCO₂Na (ii),H₃O⁺

Scheme 1.

A simple route to furo[3,2-b]pyridine depends on the one-step generation of the pyridine ring by gas-phase cyclisation of a conjugated iminyl radical (Scheme 2) (H. McNab, J.chem. Soc. Perkin I, 1980, 220. C.L. Hickson and H. McNab, Synthesis, 1981, 464).

Scheme 2.

The previously unknown 2-aminofuro[3,2-b]pyridines (33a,b,c) are prepared from 3-hydroxypyridine 1-oxide as shown below (M.L. Stein, F. Manna and C.C. Lombardi, J. heterocyclic Chem., 1978, 15, 1411. N. Cali *et al*., Gazz., 1984, 114, 211).

(32)

(33) (a) X=CO₂Et,m.p.190-1°
     (b) X=COMe,m.p. 200-1°
     (c) X=CN, m.p. 215-7°

Compound (33a) rearranges with sodium ethoxide in ethanol to give the cyclic lactone (35) via the isolable intermediate (34) (N. Desideri, F. Manna and M.L. Stein, J. heterocyclic Chem ., 1981, 18, 1085).

(34)   (35) m.p. 316-317°

Bromination of furo[2,3-b]-, furo[3,2-b]-, furo[2,3-c]-, and furo[3,2-c]pyridine gives the corresponding *trans*-2,3-dibromo-2,3-dihydro- derivatives which are converted into 3-bromo-furopyridines by treatment with sodium hydroxide in methanol. The nitration of these compounds using fuming nitric acid in concentrated sulphuric acid gives several products. The nitration of furo[2,3-b]pyridine gives a mixture of addition products and the 2- nitro derivative. Furo[2,3-b]pyridine gives a mixture of the *cis* and the *trans*-2-nitro-3-hydroxy-2,3-dihydro derivatives and the 2-nitro derivative. Nitra-tion of furo[2,3-c]pyridine gives compound (36) whilst furo-[3,2-c]pyridine gives a mixture of the 2-nitro derivative and compound (37) (S. Shiotani *et al.*, J. heterocyclic Chem., 1984, 21, 725. c.f. J.W. McFarland *et al.*, *ibid.*, 1975, 12, 705, and S. Clementi *et al.*, J. chem.Soc.Perkin II, 1978, 861).

(36)   (37)

Of the furopyridines, the [3,4-c] isomers (39) have been the least studied. The parent compound (39) is obtained by flash vacuum pyrolysis of compound (38). It melts below room temperature and polymerises to a viscous mass (W.E. Wiersum *et al.*, Tetrahedron Letters, 1977, 1741).

(38) → (39)

600–650°
0.1 Torr

The furo[3,4-c]pyridine-4-ones (41) are more stable and are prepared from 3,4-disubstituted furans (e.g. 40) by treatment with amines in methanol (D. Giardina *et al.*, J. heterocyclic Chem., 1978, 15, 993).

RNH$_2$,MeOH
0°,25d.

(40) → (41) (a) R = H
                (b) R = Me

(c) *Furoquinolines*

The main interest in furoquinolines centres around their widespread occurrance as alkaloides (M.F. Grundon in "The Alkaloids", ed. R.H.F. Manske and R.G.A. Rodrigo, Academic Press, New York, 1979, Vol.17, p.105).
The reaction of 3-haloquinolin-2-ones with copper (I) acetylide to give 3-isopropenylfuroquinolines (ca. 75% yield) is a key step in the synthesis of dihydroisopropylfuroquinoline alkaloids. Thus (42; R=H) with Cu C≡CCMe=CH$_2$ in refluxing pyridine yields (43; 75%) and similarly (42; R=H) with Cu C≡CCR^{1}Me$_2$ (R^{1}tetrahydropyranyloxy) followed by deprotection (HCl/H$_2$O) gives (44; 45%). The reaction of (42;R=Me) with Cu C≡CCMe=CH$_2$ produces (45; 22%) (J.L. Gaston, R.J. Greer and M.F. Grundon, J. chem.Res.S, 1985, 135).

The furanoquinolines (47) have been made by treatment of
the 3-2^1-chloroethylquinolin-2-ones (46) with potassium hydroxide in methanol (O.Meth-Cohn *et al.*, J. chem. Soc.Perkin
I, 1981, 2509).

(46) R = Me,OMe

(47) R = Me,m.p. 139-140°
OMe,m.p. 130°

A wide variety of furo[2,3-b]quinolines can be obtained from
3-vinylquinolin-2-ones by acetoxycyclisation using iodine in
the presence of silver acetate followed by dehydroacetoxy-
lation with phosphoric acid (P. Shanmugam *et al.*, Monatsh.,
1976, 107, 259).

(yields 83 - 100%)

Another route to furo[2,3-b]quinolines uses N-acrylaconamides
(48) as starting materials. These anilides readily undergo
intramolecular cyclisation when heated with polyphosphoric
acid (PPA) to give 1,2-dihydro-2-oxoquinoline-3-acetic acids
which can be then converted into the furoquinolines (P.
Shanmugam, *ibid.*, 1977, 108, 725).

(48)

(yields 18-57%)

The 2,3-dihydro-4-methyl-3-oxofuro[2,3-b]quinoline (49)
has been made by the route shown below. It is readily converted into 3-chloro-(50a) or 3-acetoxy-(50b)-4-methylfuro-
[2,3-b]quinoline (R. Palaniappan *et al.*, Indian J. chem.,
1981, 20B, 798).

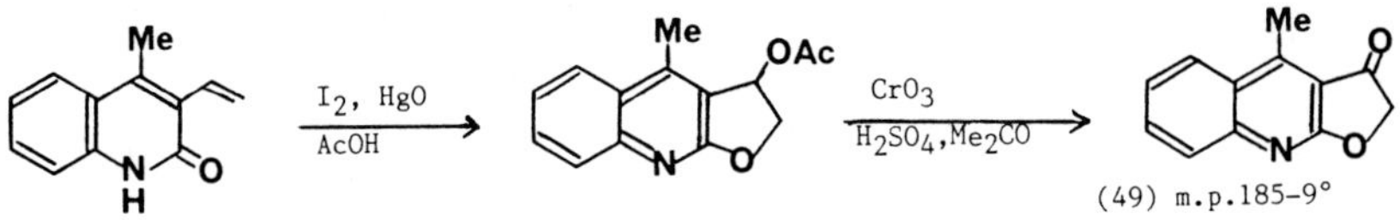

(49) m.p.185-9°

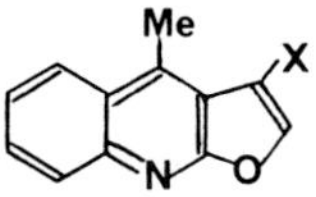

X = Cl, m.p. 156.5-157.5°
    OAc, m.p. 191-192°

(50)

An interesting observation is that the 2-quinolone (51) in
dimethylsulphoxide slowly equilibrates to the furo[2,3-b]-
quinoline (52) at room temperature (G.B. Bennet, R.B. Mason
and M.J. Shapiro, J. org. Chem., 1978, 43, 4383).

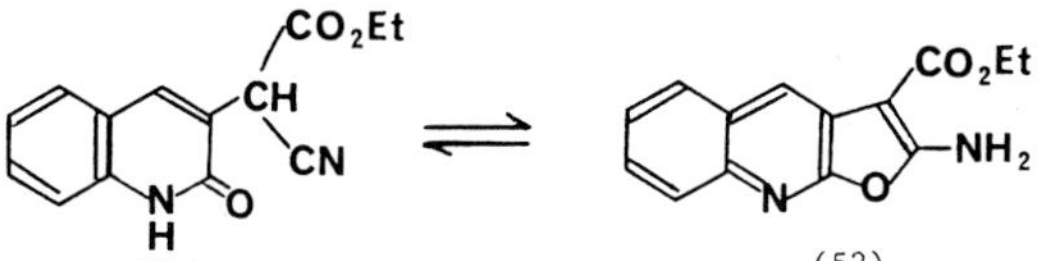

(51)   (52)

A number of methoxylated furo[2,3-b]quinolines are found as
alkaloids, particularly in *Rutaceae*. Compounds of this type
which have been isolated or synthesised are given below:
Glycarpine (53), m.p. 171°, and evolitrine (54) (which had
previously been isolated from *Evodia littoralis Eudl.)* from
Glycosmis cyanocarpa (M. Sarkar, S. Kundu and D.P.Chakra-
borty, Phytochem., 1978, 17, 2145), kokusaginine (55), m.p.
182°, (and evolitrine) from *Acryonychia pendunculata* (L.B.
de Silva *et al.*, *ibid.*, 1979, 18, 1255), and six dihydrofuro-
quinoline quaternary bases (56a-e, 57a, b) have been isolated
from *Choisya ternata, Ptelea trifoliata* and *Ruta graveolens*
(M. Rideau, *ibid.*, p.155).

(53)   (54)   (55)

(56)(a) R=8-MeO;(-) (S) balfourdinium
    (b) R=H; (-) (S) platydesminium
    (c) R=H; (+) (R)   "
    (d) R=7,8-OCH₂O; (-) (S) hydroxyluminium
    (e) R=6-OH; (+) (R) ribalinium

(57) (a) Rs = 6.8-di-OMe
         (+) ptelefolonium
     (b) Rs = 7,8-di-OMe
         (+) isoptelefolonium

(d)   *Pyranopyridines*

The thermal rearrangement of 4-pyridylpropargyl ethers gives
a mixture of cyclobutapyridines. However, 3-pyridylpropargyl
ether behaves completely differently giving in *n*-decane 2H-
pyrano[3,2-b]pyridine (59; 51%) together with 2-methylfuro-
[3,2-b]pyridine (60) and 2-methylfuro[2,3-c]pyridine (61).
In dimethylformamide only furopyridines are formed.
The 2-methylpyridylpropargyl ether (62) gives 8-methyl-2H-
pyrano[2,3-c]pyridine (63; 55%) and 2,7-dimethylfuro[2,3-c]-
pyridine (64; 28.5%) in *n*-decane but only the furopyridine
(64) in dimethylformamide (J. Bruhn *et al.*, Helv.,1978, 61,
2542).

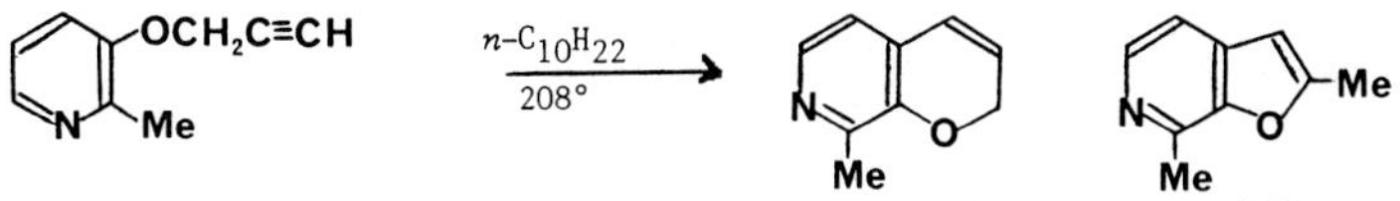

(62) (63) (64)

5-Azachroman(2,3-dihydro-4H-pyrano[3,2-b]pyridine)(65) and 2H-pyrano[3,2-b]pyridine (66) have been obtained as shown below (H. Sliwa and K.P. Krings, Heterocycles, 1979, 12. 493).

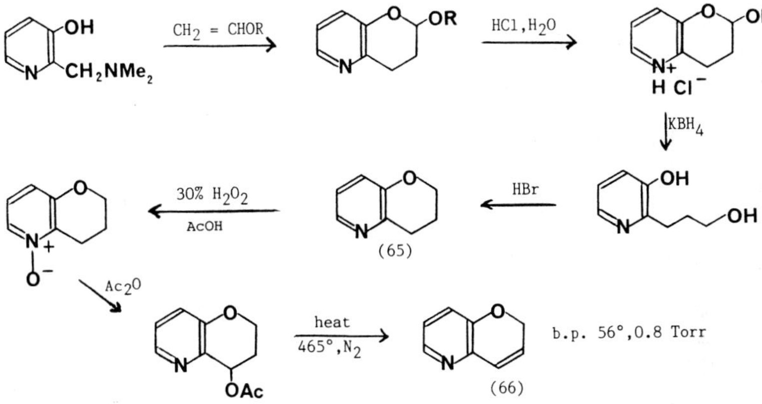

(65)

(66) b.p. 56°,0.8 Torr

Interaction of the lithium derivative of (65) with the diethyl acetal of bromoacetaldehyde followed by acid hydrolysis and subsequent ring closure gives the parent 3,4-dihydropyrano[2,-3,4-hi]indolizine (67a). The 2-phenyl analogue (67b) is obtained using phenacyl bromide (H. Sliwa, D. Blondeau and R. Rydzkowski, J. heterocyclic Chem., 1983, 20, 1613).

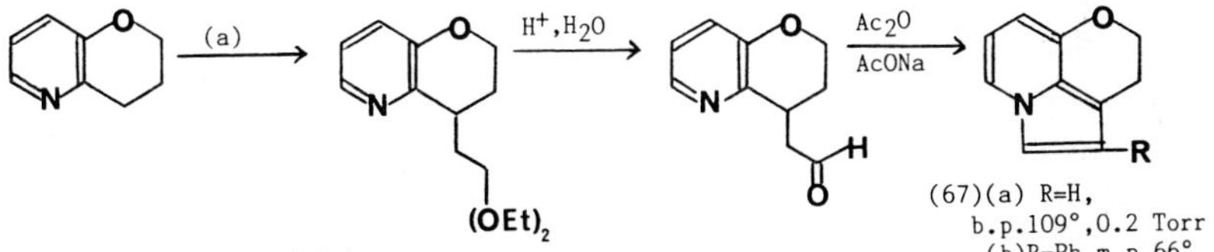

(a)(i) BuLi(ii),BrCH$_2$CH(oEt)$_2$

(67)(a) R=H,
b.p.109°,0.2 Torr
(b)R=Ph,m.p.66°

(e) *Benzopyranopyridines*

5,6-Dihydro-2H-pyran-3(4H)-one (68) reacts regioselectively
in a few cases with *ortho*-substituted phenyl carbonyl compounds to give pyran-annelated heterocycles, for example
pyrano[3,4-b]- (69) and -[3,2-b]- (70) -quinolines can be
obtained. Better results are obtained using the enamine (71)
the silylenol ether (72) or the lithium salt derived from
(68). In this way 3,4-dihydro-1H-pyrano[3,4-b]quinolin-5-
(10H)-one (73a; 28%, m.p. 320-330°), 3,4-dihydro-10-phenyl-
1H-pyrano[3,4-b]quinolin-5(10H)-one (73b; 22%, m.p. 254-257°)
and 3,4-dihydro-2H-pyrano[3,2-d]quinolin-10(5H)-one (74; 17%,
m.p. 275-278°) can be obtained (F. Eiden and K.T. Wanner,
Ann., 1984, 1759).

(68)

(69)(a) R=H,m.p.114-115°
(b) R=Me,m.p.113-114°

(70)(a)R=H,m.p.71-73°
(b)R=Me,m.p.240°
(HCl salt)
(c)R=Ph,m.p.174-176°

(71)   (72)   (73) (a) R=H
(b) R=Ph
(74)

Ethoxymethylenemalononitrile is a useful synthon for fused
heterocyclic systems. It reacts with 4-hydroxy-2-quinolones
(75a,b) to give the pyranoquinolines (76a-c). With 4-hydroxy-
coumarin (77) or its sodium salt the [1]benzopyranopyridines
(78a,b) are obtained. Further transformations of (78a) and
(78b) give the compounds (78c,d) and (79a-c) (H.-W. Schmidt,
R.Schipfer and H. Junek, Ann., 1983, 695).

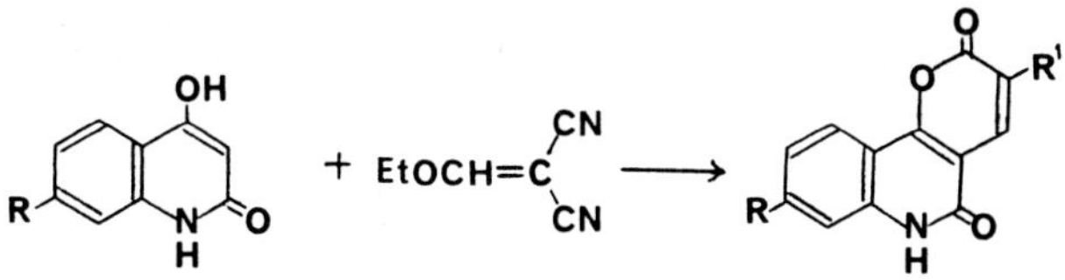

(75)(a) R=NMe$_2$, (b) R=OMe

(76)(a) R=NMe$_2$,R^1=CN (b) R=OMe,R^1 = CN

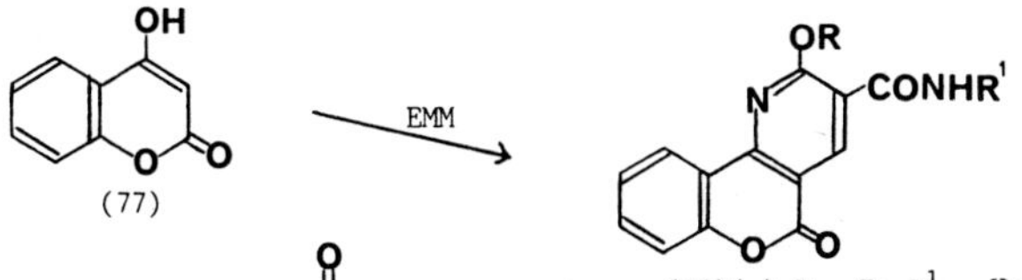

(77)

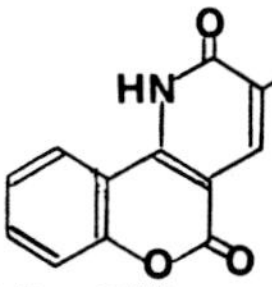

EMM

(78)(a) R = Et,R^1 = H; ᴸ.p. 270°
(b) R = Me,R^1 = H, m.p. 272°
(c) R = Et,R^1 = COMe, m.p. 228°
(d) R = Me,R^1 = COMe, m.p.216°

(79)(a) R = CONH$_2$, m.p. 346°
(b) R = CO$_2$H, m.p. 300°
(c) R = H,   m.p.>300°

A number of 1,2,3,4-tetrahydro-5H-[1]benzopyrano[3,4-c]pyri-din-5-ones (81) have been prepared (as potential bronchodi-lators or as intermediates in the synthesis of such compounds) by the Pechmann condensation of 3-carbomethoxy-1-methyl-4-piperidone (80a) with phenols. The parent compound (81) is obtained by the route shown below (D.T. Connor et al., J. heterocyclic Chem., 1984, 21, 1557).

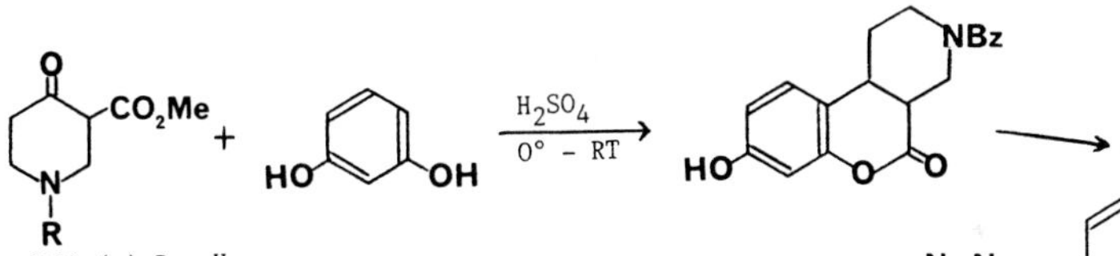

(80) (a) R = H
(b) R = CH$_2$Ph

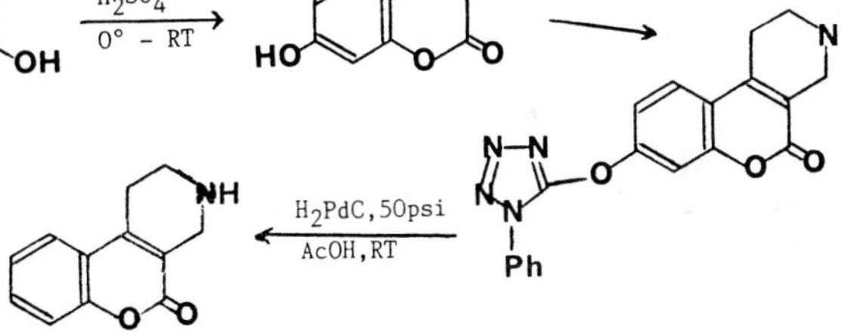

H$_2$SO$_4$
0° – RT

H$_2$PdC,50psi
AcOH,RT

(81)

Another route to benzopyranopyridines is the reaction of 2-alkoxydihydropyrans (82) with hydroxylamine hydrochloride. Both 5H-benzopyrano[1][4,3-b]pyridine (83a) and the 4-phenyl derivative (83b), and 5H–benzopyrano[1][3,4–c]pyridine (84a) and the 4–phenyl derivative (84b) have been obtained (M.–C. Bellassoued–Fargeau and P. Maitte, J. heterocyclic Chem., 1984, 21, 1549).

(82)

(83)(a) R=H,b.p.110°,0.1Torr
    (b) R=Ph, b.p.96°,0.05Torr

(84)(a) R=H,m.p.102°
    (b) R=Ph,m.p.158°

(f) *Pyrrolopyridines or azaindoles*

The Madelung synthesis for the preparation of indoles has been successfully extended to amidopyridines giving a convenient synthesis of pyrrolopyridines. Thus 2–benzamido–3–methylpyridine (85) on treatment with lithium diethylamide gives 2–phenylpyrrolo[2,3–b]pyridine (86), m.p. 203–204° in 22% yield (W.J. Houlihan, V.A. Parrino and Y. Uike, J. org. Chem., 1981, 46, 4511)

(85)

LDA,THF
−20°

(86)

The enaminone (87) from ethyl acetoacetate and *t*–butyl 2–aminocyanoacetate reacts with ethyl acrylate under basic conditions to give the pyrrolopyridine (88) which is converted into the dihydro derivative (89). The latter is aromatised to (90) by mild oxidation (T. Murata, T. Sugawara and K. Ukawa, Chem. pharm. Bull., 1978, 26, 3080).

(a) $CH_2=CHCO_2Et$, NaOEt   (b) HCO_2Et or heat   (c) DDQ or chloranil

A synthesis of pyrrolo[3,4-c]pyridines from a 3-bromopyridine bearing a C-N-C substituent in the 5-position under strongly basic conditions (e.g. potassium amide) involves a carbanion generated at the side chain cyclising on to the pyridyne formed by elimination of HBr from the pyridine (I. Ahmed, G. W.H. Cheeseman and B. Jacques, Tetrahedron, 1979, 35, 1145). The reaction and some of the products formed are given below.

(91) R=H,CO_2Et,COPh

Convenient syntheses of 4-(94) and 6-(95)azaindole-2(3H)-one involve the hydrogenolysis, decarboxylation and cyclisation of the pyridines (92) and (93) respectively (R.W. Daisley and J.R. Hambali, Synth. Comm., 1981, 11, 743).

(92) $\xrightarrow{(i),(ii)}$ (94) m.p. 204–206°

(i) H_2,Pd/c,EtOH,RT
(ii) EtOH,70°

(93) $\longrightarrow$ (95) m.p. 232°

The reductive cyclisation of the pyridine (96) gives the 1-hydroxy derivative (97). Similarly prepared is (98), and some further examples of this type of reaction have been described (R.W. Daisley and J.R. Hambali, J. heterocyclic Chem., 1983, 20, 999).

(96) $\longrightarrow$ (97) m.p. 194–196° (98) m.p. 234–236°

A one-pot synthesis of the 4-azaindole ring system involves the photo-reaction of enolate anions with 3-amino-2-chloro-pyridine. The reaction involves an $S_{RN}1$ mechanism and is exemplified by the preparation of (99a, b and c) (R. Beugel-mans, R. Boudet and L.Quintero, Tetrahedron Letters, 1980, 21, 1943).

$\xrightarrow[h\upsilon\ 350\ mm]{CH_2=\overset{O^-}{\overset{|}{C}}-R,\ NH_3}$

(99)(a) R=Me,m.p.195°
(b) R=i-C$_3$H$_7$,m.p.207°
(c) R=t-C$_4$H$_9$,m.p.240°
(subl.)

The cyclisation, under basic conditions, of the adducts of maleimide and its derivatives with 3-aminocrotonates (or 3-aminocrotonitrile) gives pyrrolo[3,4-c]pyridines and not pyrrolo[2,3-b]pyrroles as had been previously reported (K.R. Shah and C.D. Blanton, J. org. Chem., 1982, 47, 502).

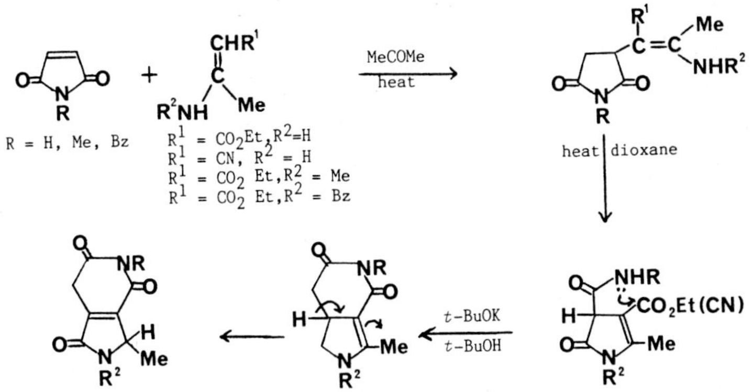

An unusual heterocyclic rearrangement is seen when 1-alkyl-4-chloro-1H-pyrrolo[3,2-c]pyridines react with an excess of primary alkylamine. In addition to the expected 4-alkylamino-pyrrolo[3,2-c]pyridines alkylaminopyrrolo[2,3-b]pyridines are also obtained resulting, as illustrated below, from a reversible isomerisation of the preceding compounds (E. Bisagni, M. Legraverend and J.-M. Lhoste, J.org.Chem., 1982, 47, 1500).

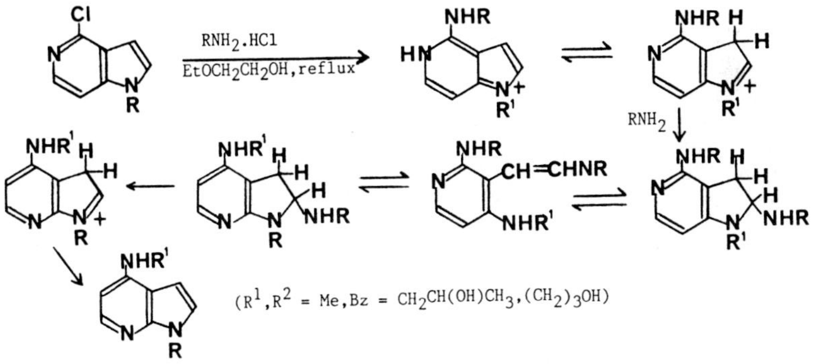

$(R^1, R^2 = Me, Bz = CH_2CH(OH)CH_3, (CH_2)_3OH)$

The acylation reaction of 1H-pyrrolo[2,3-b]pyridine with acetic anhydride, acetic acid gives 1-acetylpyrrolo[2,3-b]-pyridine (100) in >90% yield. By using acetic anhydride in the presence of aluminium chloride >90% of 3-acetyl-1H-pyrrolo[2,3-b]pyridine (101) is obtained.
2-Chloroacetyl chloride in the presence of aluminium chloride also gives the 3-substituted product (C. Gálvez and P. Viladoms, J. heterocyclic Chem., 1982, 19, 665).

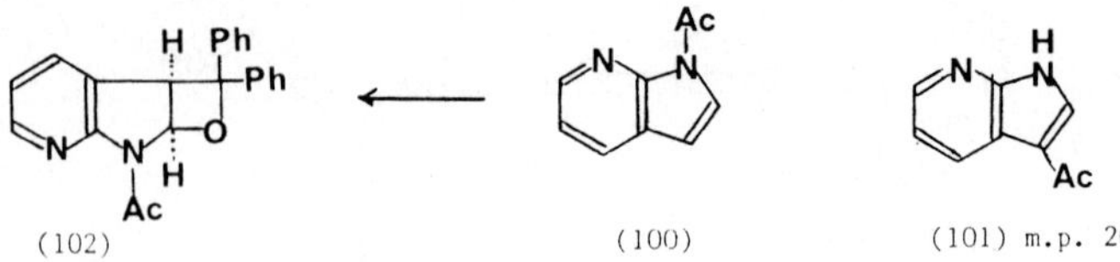

(102)   (100)   (101) m.p. 206-207°

It is found that 1-acetylpyrrolo[2,3-b]pyridine (100) reacts
with benzophenone on irradiation to give (102) whereas the
unsubstituted system fails to react (T. Nakano and M. Santana,
J. heterocyclic Chem., 1976, 13, 585).
The synthesis of 4-aminopyrrolo[3,2-c]pyridine followed by
its ribosylation does not give the desired deazatubercidin
but the isomer (103)(88%, m.p. 134-138°) (C. Ducrocq *et al.*,
Tetrahedron, 1976, 32, 773).

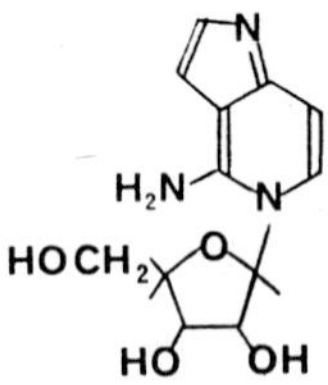

(103)

The aza-2-pyrrolo[1,2-a]indole derivative (105) is formed by
a 1,3-dipolar cycloaddition of dimethyl acetylenedicarboxylate with the mesoionic intermediate (104) followed by the
sequence of reactions shown below (D. Laduree, J.-C.Lancelot
and M. Robba, Tetrahedron Letters, 1985, 26, 1295). The
reaction of (105) with phosphorus oxychloride gave the chloro
derivative (106).

(106) m.p. 162°   (105) m.p. > 260°

(i) MeNO$_2$, base (ii) reduction (iii) Et$_3$N

The azapyrroloindoles (107 – 110) are obtained by the reactions shown in Scheme 3 (D. Laduree and M. Robba, Heterocycles, 1984, 22, 303).

(a) NaH,Cu$_2$Cl$_2$,DMSO (b) LiAlH$_4$,AlCl$_3$ (c) NaBH$_4$,dioxane (d) NH$_2$NH$_2$, EtOH

Scheme 3

The chemistry of 1-H-pyrrolo[3,2-b]-, pyrrolo[3,4-b]-, 3-H-pyrrolo[2,3-c]- and 1-H-pyrrolo[3,2-c]quinolines has been reviewed (M.A. Khan and J.F. da Rocha, Heterocycles, 1977, 6, 1978, 9, 1059, 1978, 9, 1617, 1979, 12, 857).
The o-nitrophenyldipyrrolylmethane (111, R=NO$_2$) is reduced by triethylphosphite to the nitrene (111,R=N) which forms the pyrrolo[3,2-b]quinoline (112) by an intramolecular insertion reaction (G. Jones and W.H. McKinley, J.chem.Soc. Perkin I, 1979, 599).

(111)   (112)

The ring closure and rearrangement of 3-(*N*-2-chloroallyl)-
quinolines, catalysed by Lewis acids gives pyrrolo[3,2-c]-
quinolines, e.g. (113) (B.G. McDonald and G.R. Proctor, J.
chem.Soc. Perkin I, 1975, 1446).

(113) m.p. 158-160°

Pyrroloquinolines can be obtained by Friedländer condensation
between 2-aminobenzaldehyde and pyrrolidinones. Such a
reaction using the pyrrolidin-3-one (114) gives different
products depending on the reaction conditions. In base the
pyrrolo[3,2-b]quinoline (115), and its hydrolysis product,
represent 85% of the yield and the pyrrolo[3,4-b]quinoline
(116) the other 15%. These figures probably reflect the
relative stabilities of the carbanions derived from the
ketone (114). Under acid catalysis only (116) is obtained
in 88% yield. Here the first formed imine would ring close
to minimise steric interactions. Thus this procedure
enables either type of ring system to be obtained and the
reaction can also be carried out using aminopyridine aldehydes (L.H. Zalkow *et al.*, J.chem.Soc. C, 1971, 3551).
This approach has been extended to give aza analogues of
ellipticene (F.Nivoliers *et al.*, Tetrahedron Letters, 1980,
21, 4485).

(114)   (115)   (116)

Phenylhydrazones such as (117) formed from pyrrole-3-alde-
hydes undergo Fischer-type reactions on heating with acid to
give the pyrroloquinolines (118) (H. Fritz and S.Schenk,
Ann., 1975, 255).

(117)

(118) R=Me,Et,Ph,Bz

(g) *Carbolines (pyridinoles)*

The α-carbolines(pyrido[2,3-b]indoles) have been reviewed
(in Russian) (A.A. Semenov and V.V. Tolstikhina, Khim.
Geterotsikl. Soedin., 1984, 435).
Thermal decomposition of the azidoacrylates (119) derived
from 2-substituted indole-3-carbaldehydes gives β-carbolines
(120) (C.J. Moody and J.G. Ward, J.chem.Soc. Perkin I, 1984,
2895). Also obtained are compounds (121) and (122) but in
the case of the formation of compounds (122) the azepino-
indoles (123) seem to be the favoured products. The use of
such azidoacrylates with 2-substitution gives rise to 1,8-
dihydropyrrolo[2,3-b]indoles (C.J. Moody and J.G. Ward,
ibid., p.2903).

(119) R^1 = Bz,CH$_2$OMe

R = Me,Et, etc

(120) R^1 = Bz, m.p. 120-121°
R^1 = CH$_2$OMe, m.p. 128.5-130.5°
R^1 = H, m.p. 224-229°

(121) m.p. 219-223°

(122) R = Me, 43%
R = Et, trace

(123)

4-Substituted β-carbolines can be prepared from indole
derivatives of type (124). Such reactions and further
metatheses including halogenation, side chain functional-
isation, nitration, etc. have lead to the synthesis of some
80 such compounds (G. Neef, *et al*., Heterocycles, 1983, 20,
1295).

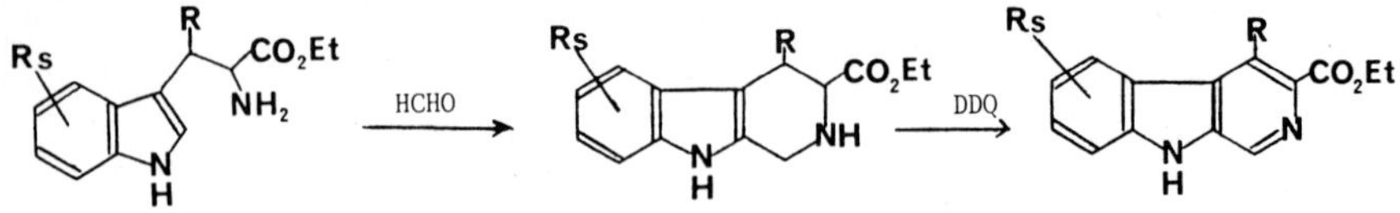

(124)

Tetrahydro-5H-pyrido[3,2-b]- and tetrahydro-9H-pyrido[2,3-b]-
indoles have been made by the reaction of enamines with
tetrachloro-4-cyanopyridine. The proportion of the two
types of product varies with the dialkylamino group, the
yields of the isolated carboline products being shown below.
The tetrahydropyridoindoles can be aromatised by treatment
with DDQ(2,3-dichloro-5,6-dicyano-*p*-benzoquinone)
(H. Suschitzky *et al*., J.chem.Soc. Perkin I, 1983, 637).

ENAMINE	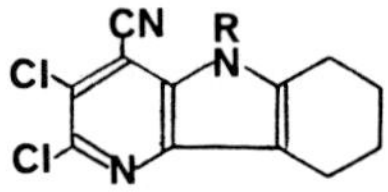	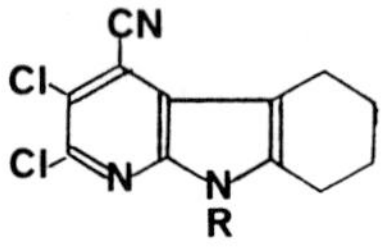	R
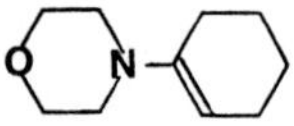	11	48	$CH_2CH_2OCH_2CH_2Cl$
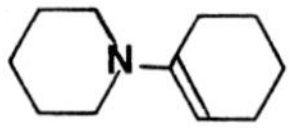	42	41	$(CH_2)_5Cl$
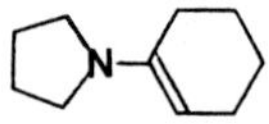	26	37	$(CH_2)_4Cl$
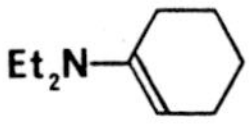	31	trace	Et

1-Ethyl-4-oxo-1,2,3,4-tetrahydro-β-carboline (125) is
converted into 1-ethyl-4-amino-β-carboline (126b) by heating
in hydrazine (M, Cain, R. Mantei and J.M. Cook, J.org.Chem.,
1982, 47, 4933). Further study of this reaction has shown
the mechanism to be that shown in Scheme 4. After removal

of the benzamide moeity and simultaneous formation of the
hydrazone, elimination of ammonia is followed by a proton
shift. (N. Fukada *et al.*, Tetrahedron Letters, 1985, <u>26</u>,2139).

Scheme 4

Fukada *et al.*, (*loc.cit.*) have reported the thermal [3,3]
sigmatropic rearrangement of the allyl ether (127) to the 3-
allyl substituted β-carboline (128).

Three convenient syntheses of γ-carbolines have been described (H. Akinmoto, A. Kawai and H. Nomura, Bull. chem.Soc.
Japan, 1985, <u>58</u>, 123).
(i)   Acid-catalysed cyclisation of 2-acetamido-3-(2-indolyl)-
alkanoic acids to 1,2-dihydro-γ-carbolines followed by dehydrogenation to the γ-carbolinecarboxylates, hydrolysis of
the ester function, and conversion to the amino compound.

(i) Hydrolysis   (ii) DPPA

(ii)   Formation of a 4-(benzotriazolyl)pyridinamine compound
then cyclisation using polyphosphoric acid.

(iii)   Cyclisation of 3-acetylindole-2-acetonitrile using
methanolic ammonia.

Condensation of 1-acetyl-3-indolinone (129) with an aldehyde
followed by cycloaddition to malononitrile gives the pyrano-
[3,2-b]indole (130). The latter rearrange and are deacetyl-
ated on treatment with alcoholic potassium hydroxide giving
the δ-carbolines (131) (V.S. Velezheva, K.V. Nevskii and
N.N. Suvarov, Khim. Geterotsikl. Soedin., 1985, 230).

(129)   (130)   (131)

The naturally occurring pyridindole (-)-trypargine (133) isolated from the skin of the African frogs *Kassina senegalensis* has been synthesised by the reaction of tryptamine with the compound (132) (M. Shimizu *et al.*, Chem. pharm. Bull., 1982, <u>30</u>, 909).

(132)

(133) m.p. 254-7°(as sulphate)

(h) *Naphthyridines (diazanaphthalenes)*

The following reviews of naphthyridines are available: W.W. Paudler and R.M. Sheets, Adv. heterocyclic Chem., 1983, <u>33</u>, 147; W. Czuba, Khim. Geterotsikl. Soedin., 1979, 3; M. Yamamoto *et al.*, Heterocycles, 1982, <u>19</u>, 1263; P.A. Lowe in "Comprehensive Heterocyclic Chemistry", Vol.2, p.581, *Pergamon Press, Oxford*, 1984. A review of the nucleophilic substitution reactions of naphthyridines and the ^{13}C-nmr spectral data of all the parent naphthyridines (in $CDCl_3$) has been published (H.C. van der Plas, M. Wozniak and H.J.W. van den Haak, Adv. heterocyclic Chem., 1983, <u>33</u>, 95). A synthesis of 1,6- and 1,7-naphthyridines starts from the readily available pyridinane (134). Sequential treatment of this compound with hydrogen peroxide in acetic acid, acetic anhydride, then concentrated sulphuric acid gives a mixture of 7H-1- (135) and 5H-1- (136) pyrindine. Ozonolysis of this mixture and treatment with ammonium hydroxide gives 1,6- (137) and 1,7- (138) naphthyridine in a ratio of 1:2. The 5-methyl and 5-benzyl-1,6-naphthyridines and 8-methyl and 8-benzyl-1,7-naphthyridines have also been prepared by this reaction (C.S. Giam and D. Ambrozich, Chem.Comm., 2984, 265).

(134)   (135)   (136)   (137)   (138)

(a)(i) 30% H_2O_2,AcOH, 75° (ii) Ac_2O,100° (iii) H_2SO_4,125° (b)O_3,Me_2S,NaHCO$_3$,NH$_4$OH

1,2,3,4-Tetrahydro-1,6-naphthyridines can be obtained from enamines. For example condensation of the enaminoamide (139a) with dimethylformamide diethylacetal (DMFDEA) gives the pyrido[1,2-c]pyrimidine (140) which rearranges on refluxing in aqueous solution to give the 1,6-naphthyridine (141a) (V.G. Granik, E.O. Sochreva and N.P. Solov'eva, Khim, Geterotsikl, Soedin., 1980, 416; V.G. Granik *et al.*, *ibid.*, 1981, 1120).

(139)(a) R=CONH$_2$
     (b) R = CN

(140)

(141)(a) R=OH
     (b) R=NH$_2$

However a related reaction to that above in which the nitrile (139b) is reacted with DMFDEA followed by ammonia gives the 1,6-naphthyridine (141b) without the intermediacy of the pyridopyrimidine (V.G. Granik, A.M. Zhidkova and R.A. Dubinskii, Khim. Geterotsikl. Soedin., 1982, 518).
Ethyl *N*-benzylacetimidate (142) and diketene react in acetic acid at room temperature to give the 1,6-naphthyridine (143) by a mechanism which has not been clarified. Further transform-ations of this product lead to other 1,6-naphthyridines, e.g. (144) and (145) (T. Kato and T. Sakamoto, Chem. pharm. Bull., 1975, 23, 2629).

(142)

(143) m.p. 117°

(144) (a) R=Bz
      (b) R=H

(145) (a) R=Cl
      (b) R=H

A further example of an enamine synthesis of naphthyridines is the reaction of 3-cyano-4-methylpyridine with DMFDEA to give the enamine (146) followed by cyclisation with acid to give the naphthyridine (147).

(146)   (147) m.p. 260-262°

The use of 3-cyano-2-methylpyridine gives 5-hydroxy-1,6-naphthyridine (148a) as well as the bromo derivative (148b). However the same reaction using the enamine intermediate (149) gives a very low yield (about 5%) of the 1,7-naphthyridine (150) (J.J. Baldwin, K. Mensler and G.S. Ponticello, J.org.Chem., 1978, 43, 4878).

(148) (a) X=OH   (149)   (150)
      (b) X=Br

A convenient synthesis of 5-hydroxy-1,7-naphthyridine (155) involves condensation of ethyl (2-bromomethyl)nicotinate (151) with N-(p-toluenesulphonyl)glycine ester (152) to give the derivative (153) which cyclises on treatment with sodium ethoxide in ethanol to give the naphthyridine (154). The latter on hydrolysis gives (155) (P. Messinger and H. Meyer, Ann., 1978, 443).

(151)   (152)   (153)   (b) (i)(ii)   (154)

(a) NaOMe,EtOH   (b)(i) NaOH,H₂O,heat (ii) HCl,H₂O heat

(155) m.p. 230°(subl.)

Heating a 2-aminopyridine (156) with diethyl ethoxymethyl-
enemalonate in "Dowtherm A" gives a product (157) which
undergoes 1 → 3, N → C acyl migration on heating at 250° to
yield a 1,8-naphthyridine (158) (Z.Meszara and I. Hermecz,
Tetrahedron Letters, 1975, 1019; I. Hermecz *et al*., J. chem.
Soc. Perkin I, 1977, 789). This reaction has also been used
to obtain 1,2,3,4-tetrahydro-1,8-naphthyridines (I. Hermecz
and Z. Meszara, Heterocycles, 1979, 12, 1407), benzo-1,8-
naphthyridines (G. Bernath *et al*., J. heterocyclic Chem.,
1979, 16, 137) and anthyridines.

(156) (a) R=Me   (157)   (158)
      (b) R=OEt
      (c) R=NH₂

Other reactions which lead to naphthyridines but which
either give mixtures of products or give low yields include
the following.
(a)   The reaction between 4-dialkylamino-3,4-dehydropiperidin-
2-ones(and thiones) (159) and bis(2,4,6-trichlorophenyl)
ethylmalonate(160) to give 1,6-naphthyridines (161) (G.
Zigeuner, K. Schweiger and D. Habernig, Monatsh.   1982,
113, 573).

(159) X=O or S   (160)   (161)

(b)   The condensation between ethyl 3-alkylaminocrotonates
(162) in the presence of 3-mercaptopropanoic acid to give
(163) (M. Yamamoto *et al*., Heterocycles, 1982, 19, 1263).

(162) R=Me,Bz   (163)

(c) The formation of tetracyclated isobutenes and their cyclisation using liquid ammonia to yield 1,3,6,8-tetra-alkyl-2,7-naphthyridines (C.H. Erre and C. Roussel, Bull. Soc.chim. France, 1984, 449; C. Erre *et al.*, Tetrahedron Letters, 1984, 25, 515).

(d) The cyclisation of the pyridine derivative (164), formed from malononitrile dimer and 2,4-diketones, with sodium methoxide in methanol, or HBr in acetic acid, gives the products (165) and (166) respectively. The latter can be transformed into (167) (G. Koitz, B. Thierricufer and H. Junek, Heterocycles, 1983, 20, 2405).

(e) Polyphenyl-2,6- and -2,7-naphthyridines are prepared by the condensation reactions of pyridines (170) [obtained from 3,4-dibenzoyl-2,5-diphenylthiophene (168) and amines RCH_2-NH_2 (R=Ph,CN,CO_2H) followed by oxidative ring cleavage of the resulting thieno[3,4-c]pyridines (169)] (S. Mataka, K. Takahashi and M. Tashiro, J. heterocyclic Chem., 1983, 20, 971).

(168)　(169)　(170)

The naphthyridine ring system is very π-electron deficient
and is therefore very susceptible to attack by nucleophiles.
1,5-, 1,6-, 1,7- and 1,8-Naphthyridines each react with
phenyllithium to give the corresponding 2-phenyl derivative
(Y. Hamada, I. Takeuchi and M. Hirota, Chem. pharm. Bull.,
1974, 22, 495); I. Takeuchi and Y. Hamada, *ibid.*, 1976,
24, 1813).
1,8-Naphthyridine reacts with butyllithium similarly to give
the 2-butyl derivative but the reaction of butyllithium
with dithiane followed by the addition of 1,8-naphthyridine
generates the bisnaphthyridine (171) (M. Weissenfels and
B.Ulrichi, Z.Chem., 1978, 18, 382).

(171)

The Reissert reaction of 1,7-naphthyridine gives the 8-cyano-
7-acyl product (172). However the yields are generally poor,
except for the case of diphenylcarbamoyl chloride, although
they are improved by using phase-transfer catalysis (Y.
Hamada and K. Shigemura, Yakugaku Zasshi, 1979, 99, 982).

(172) R=Me,Et,Pr

The reaction of 1,6-naphthyridine with cyanogen bromide in
methanol gives an N-cyano compound (173) which can be

converted into either 8-bromo-1,6-naphthyridine (174) or
7-methoxy-1,6-naphthyridine (175) (Y. Hamada, M. Sugiura and
M. Hirota, *ibid.*, 1978, 98, 1361).

(173)

(174)   (175)

Acid hydrolysis of the Reissert compound from 1,6-naphthyri-
dine, benzoyl chloride and potassium cyanide, gives 1,6-
naphthyridine-5-carboxamide (176) together with the
corresponding carboxylic acid (Y. Hamada and K. Shigemura,
ibid., 1979, 99, 1225).

(176) (a) R=NH$_2$ (177) (a) R=CH$_2$CO$_2$H
      (b) R=OH   (b) R=CH(CO$_2$Et)$_2$

1,6-Naphthyridine reacts with acetic anhydride to give (177a)
(T. Yamazaki *et al.*, J. heterocyclic Chem., 1979, 16, 527)
and it reacts with diethyl malonate in acetic anhydride to
give (177b). It has been established that such reactions
using acylation reagents occur at the "isoquinoline" type
nitrogen in the naphthyridines (H. Yamanaka, T. Shiraishui
and T. Sakamoto, Chem. pharm. Bull, 1981, 29, 1056).
Methylation of 1,5- 1,6-, 1,7- and 1,8-naphthyridines gives
N-mono methiodides. In the 1,6- and 1,7-naphthyridines it
is also the "isoquinoline" nitrogens N-6 and N-7 that are
methylated, this being confirmed by ^{13}C-nmr spectroscopy
(P. van de Weijer, C. Mohan and D.M.W. van den Ham, Org.mag.
Res., 1977, 10, 165).
The amination of 1,5-naphthyridine using potassium amide in
liquid ammonia at -40° gives the σ-adduct formed by the

addition of NH_2 to the 2-position. However raising the
temperature to 10° results in its conversion into a 4-σ-
adduct. Raising the reaction temperature to 50° gives the
corresponding amino substitution products (H.J.W. van den
Haak, H.C. van der Plas and B. van Veldhuizen, J.org.Chem.,
1981, 46, 2134).

In the case of 1,7-naphthyridine a mixture of 2- and 8-
adducts is formed at -10°, a result which agrees with the
electron density calculations. However on amination at room
temperature only the 8-amino product is obtained. Both 2,6-
and 2,7-naphthyridine give σ-adducts at -40°, with addition
to the 1-position, which form the 1-amino products at higher
temperatures (10° - 50°) (H.J.W. van den Haak, H.C. van der
Plas and B. van Veldhuizen, J. heterocyclic Chem., 1981, 18,
1349; M. Wozniak et al., Recl. J.R. Neth. chem.Soc., 1983,
102, 359).
The amination of 3-nitro-1,8-naphthyridines (177) gives the
corresponding 4-amino derivative (178) via 4-amino-1,4-
dihydro compounds (M. Wozniak, H.C. van der Plas and B. van
Veldhuizen, J. heterocyclic Chem., 1983, 20, 9). In the
case of 3,6-dinitro-1,8-naphthyridines these compounds give
readily isolated 4-amino-1,4-dihydro σ-adducts (179) (M.
Wozniak et al., ibid., 1985, 22, 761).

(177) R=H,Cl,NH$_2$,OEt   (178)   (179)

The reaction of 1,5- and 1,8-naphthyridine N-methylazinium
salts with liquid ammonia and potassium permanganate gives
the products shown below (M. Wozniak, D.J. Buurman and H.C.
van der Plas. J. heterocyclic Chem., 1985, 22, 765).

This same treatment of 7-methyl-1,7-naphthyridinium salt (180) gives the 8-imino compound (181) and two ring contraction products, 3-amino-1,3-dihydro-2-methyl-4-azaisoindol-1-one (182) and 1,3-dihydro-2-methyl-4-azaisoindol-1,3-dione (183). Similarly 6-methyl-1,6-naphthyridinium salt (184) gives the 5-imino compound (185). ^{1}H-nmr spectroscopy unequivocably shows that the compounds (180) and (184) are converted with ammonia into 8-amino-7,8-dihydro-7-methyl-1,7-naphthyridine (186) and 5-amino-5,6-dihydro-6-methyl-1,6-naphthyridine (187) respectively (M. Wozniak and H.C. van der Plas, J. org. Chem., 1985, 50, 3435).

(180)   (181)   (182)   (183)

(184)   (185)   (186)   (187)

The N-amino derivatives of the naphthyridines (e.g. 188,189) react with dimethyl acetylenedicarboxylate to give tricyclic compounds of the type (190) and (191) (Y. Tamura, Y. Niki and M. Ikeda, J. heterocyclic Chem., 1975, 12, 119). Treatment of these same N-aminonaphthyridines with base forms N-ylides which then give dimeric products (192). They also undergo ring expansion on irradiation to give, for example, (193) (T. Tsuchiya, M. Enkaku and H. Sawanishi, Heterocycles, 1978, 9, 621; T. Tsuchiya et al., Chem.pharm. Bull., 1979, 27, 2183)

A bromination procedure for naphthyridines involving treatment of the naphthyridine hydrohalide with bromine in nitrobenzene has been applied to the 1,7- and 1,8- isomers. In the case of the 1,7- isomer (194), as its hydrobromide, such treatment using 1.1 equivalents of bromine gives (195), (196) and (197) in the yields shown. The use of excess bromine (2.5 equivalents) gives almost exclusively the dibromo product (197) in 75% yield. A significant difference in the product ratio is observed when the hydrochloride salt is brominated using 1.1 equivalents of bromine; the products are contaminated with some chloronaphthyridines. Bromination of 1,8-naphthyridine hydrobromide (198) by this method gives a 1.1 ratio of compounds (199) and (200) when 1.1 equivalents of bromine are used. Increasing the amount of bromine to 2.5 equivalents gives the dibromo product (200) in 73% yield (H.C. van der Plas and M. Wozniak, J. heterocyclic Chem., 1976, 13, 961).

(194)	(195)	(196)	(197)
X = Br	13.5%	8.1%	45.6%
Cl	1.5%	30.9%	6.4%

| | 1.1 equivs. Br$_2$ | 32% | 30% |
| | 2.5 " Br$_2$ | >1% | 75% |

(198)　　　　　　　　　　　　　　(199)　　　　　(200)

Enoxacin, a potent antibacterial agent, is 1-ethyl-6-fluoro-1,4-dihydro-4-oxo-7-(1-piperazinyl)-1,8-naphthyridine-3-carboxylic acid (201). A satisfactory synthesis involves the preparation of a 2-substituted-6-acetylamino-3-fluoropyridine by a Balz–Schiemann reaction of the corresponding 3-amino-pyridine then elaboration of the second pyridine ring by a standard procedure (J. Matsumoto *et al*., J. heterocyclic Chem., 1984, **21**, 737).

(201)

(i) *Benzonaphthyridines (diaza-anthracenes and -phenanthrenes) and related compounds*

Cyclisation of the 2-aminonicotinic acid derivatives of the type (202) with sulphuric acid or polyphosphoric acid gives the benzo-[b][1,8]-naphthyridines (203). Similar treatment of the 3-cyanopyridines (204) gives the 10-amino derivative (205) (A.I. Mikhalev and M.E. Konshin, Khim., Geterotsikl. Soedin., 1977, 1241; N.I. Shramm and M.E. Konshin, *ibid*., 1982, 674).

(202)　　　　　　　(203)　　　　　　(204)　　　　　(205)

However under the acidic conditions some hydrolysis of
the 1-amino derivatives (205) also occurs to give the benzo-
naphthyridid-10-ones.
A general route leading to benzonaphthyridines is the
Friedlander synthesis which involves the base catalysed
condensation of o-aminoformylquinolines with compounds of the
type RCH_2COR^1. Some of the many compounds made by this
route are referred to below:
Benzo[f][1,7]naphthyridines (206) (H.E. Baumgarten et al.,
J. heterocyclic Chem., 1981, 18, 925; A. Godard et al.,
Compt. Rend., 1977, 284C, 459), benzo[h][1,6]naphthyridines
(207) (A. Godard et al., $loc.cit$.) benzo [b][1,5-naphthyri-
dines (208), benzo[b][1,8]naphthyridines (209) and also
benzo[f][1,7]naphthyridines (A. Godard and G. Queguiner,
J. heterocyclic Chem., 1982, 19, 1289).

(206)   (207)   (208)   (209)

The acid catalysed cyclisation of 2-aminonicotinaldehyde (210)
with cyclohezanone gives cyclohexa[b][1,8]naphthyridine (211)
(R.P. Thummel and D.K. Kohli, J. heterocyclic Chem., 1977,
14, 685).

(210)   (211)

The reaction between acetanilide and dimethylformamide/
phosphorus oxychloride under Vilsmeier conditions gives a
1,9-diazaanthracen-2-one (213) via the intermediate (212)
(O.Meth-Cohn and B. Tarnowski, Tetrahedron Letters, 1980,
21, 3721).

(212)   (213)

3-Methyl-2,10-diazaanthracen-1-one (216) has been obtained
by two methods from the pyridine derivative (214). In one
the aniline derivative (215) is produced which on treatment
with DMF/POCl$_3$ gives the product (C. Rivalle and E. Bisagni,
J. heterocyclic Chem., 1980, 17, 245) whilst in the other
the pyridine hemiacetal (217) is formed which on heating
generates the quinone methiodide (218). The latter reacts
with aniline to give (216) (J.L. Asherson, O. Bilgic and
D.W. Young, J. chem. Soc. Perkin I, 1980, 522).

(215)   (216)

(214)

(217)   (218)

(a) Et$_2$NCH$_2$CH$_2$COMe, MeI, KOH

The parent 3,9-diazaphenanthrene(benzo[b][1,4]naphthyridine)
(220) is obtained by cyclising the 4-anilinomethyl-3-bromo-
pyridine (219) using potassium amide in liquid ammonia
followed by oxidation of the dihydro intermediate (S.V.
Kessar et al., Proc. Indian Acad.Sci., 1979, 88A, 191).
Similar treatment of the secondary amine (221) gives (223a)
via (222) (S.V. Kessar et al., Tetrahedron Letters, 1976,
3207). Debenzylation of (223a) gives the alkaloid perlolidine (223b).

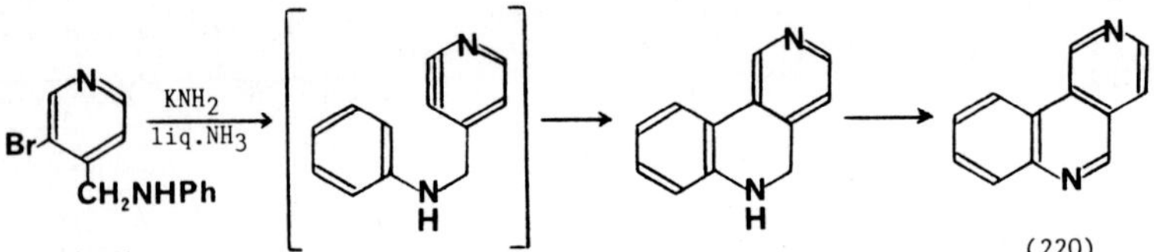

(219) (220)

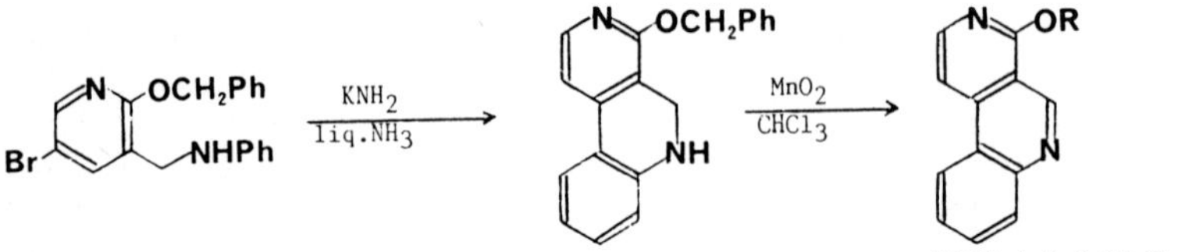

(221) (222) (223) (a) R=CH₂Ph
(b) R=H

Perlolidine (223b) may also be synthesised (81% yield) by the thermally induced Curtius type rearrangement of the quinoyl-acrylic acid (224). Similarly, the azide (225) gives the perlolidine isomer (226) (I. Lalezari and S. Nabahi, J. heterocyclic Chem., 1980, 17, 1761).

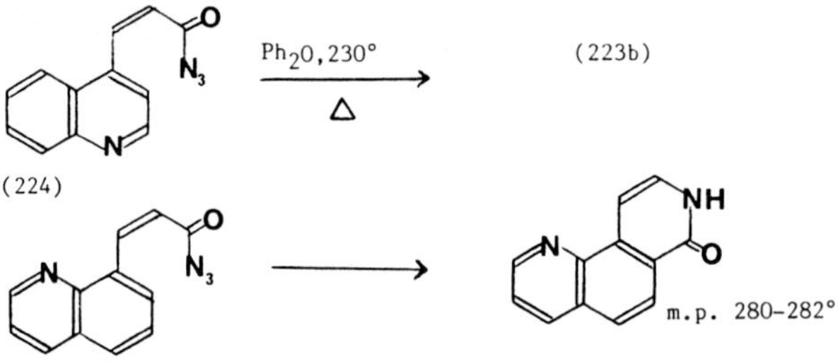

(224) (223b)

(225) (226) m.p. 280-282°

The reaction of the isoquinoline (227) with ethyl aceto-acetate under conditions of the Knorr reaction (hot poly-phosphoric acid) gives the benzo[c][1,8]naphthyridinone (228). However under less acidic conditions (hot acetic acid) the product (229) is obtained (L.W. Deady, Austral. J.Chem., 1984, 37, 1135).

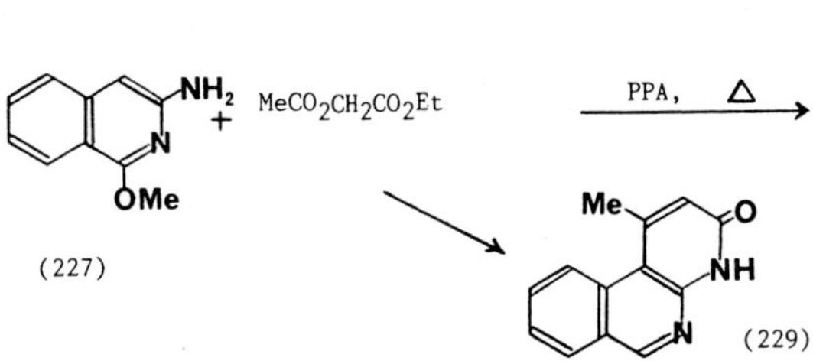

(227)

(228)

(229)

Dibenzo[c,h][1,6]naphthyridine (231) can be synthesised by
a reaction similar to that used to obtain benzo-2,6- and
benzo-3,6-naphthyridines via a benzyne intermediate from the
Schiff's base (230). The dibenzo[c,h][2,6]naphthyridine
(232) can be obtained similarly (S.V. Kessar *et al.*, Ind.
J. Chem., 1978, 16B, 92).

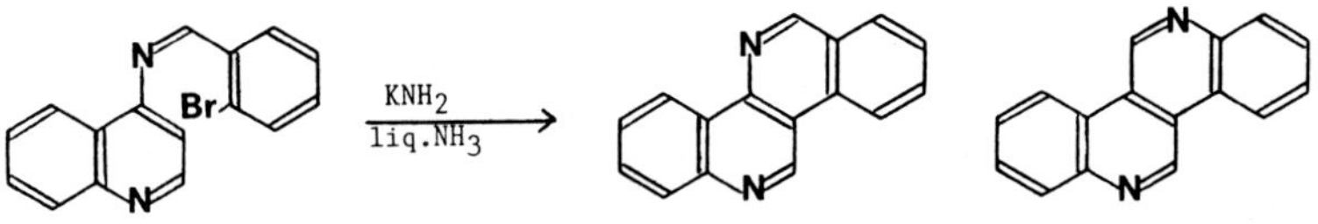

(230)

(231)

(232)

Thermolysis of the dihydro-1,4,2-oxazaphosphol-4-enes (233)
gives the nitrile ylides which dimerise to the dibenzo[c,h]-
[1,5]naphthyridine derivatives (234) (K. Burger *et al.*,
Z. Naturforsch., 1981, 36B, 345).

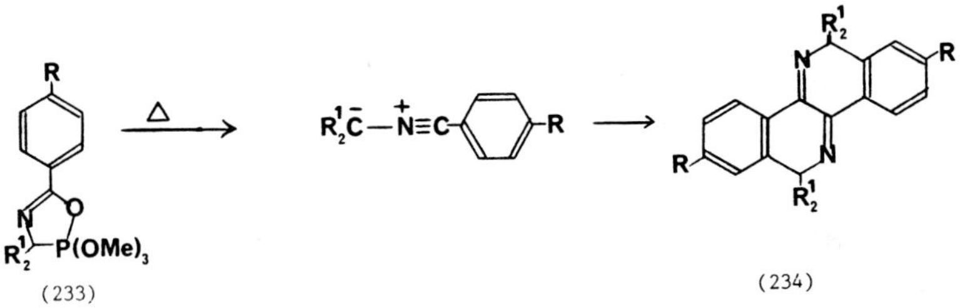

(233)

(234)

Other syntheses of this ring system have been reported
(R. Oels, R. Storer and D.W. Young, J. chem.Soc. Perkin I,
1977, 2546; J.L. Asherson and D.W. Young, *ibid.*, 1980, 512;
W. Stadlbauer and T. Kappe, Monatsh ., 1982, 113, 751).

The chemistry of the diazaphenalene ring system which can be considered as a benzonaphthyridine system, has been reviewed (S.J. Lee and J.M. Cook, Heterocycles, 1983, 20, 87). The first reported member of this system, 1,6-diazaphenalene (235), is obtained by the reaction sequence below (M.I. El-Sheikh, J.C. Chang and J.M. Cook, Heterocycles, 1978, 9, 1561; 1979, 12, 903).

(235)

(a)(i) H_2O,pH5.4   (ii) NH_3,MeOH   (b) NH_2OH, EtOH   (c) CF_3CO_2H (d) $POCl_3$

Bromination of 1,6-diazaphenalene gives mainly 2,3-dibromo- and 2,3,4-tribromo-1,6-diazaphenalene. Halogenation of the conjugate acid gives the 7-isomer (237) as the major product. Other reactions of the conjugate acid (236) also give the 7-(substituted-1,6-diazaphenalenes (237) (S.J. Lee and J.M. Cook, Heterocycles, 1981, 16, 125; R.W. Weber et al., Canad. J. Chem., 1982, 60, 3049).

(236)   (237)

(a) NBS,CF_3CO_2H → X = Br,65%   (d) $NaNO_2$,CF_3CO_2H(-60°) → X = NO_2, 70%
(b) NIS,CF_3CO_2H → X = I, 50%   (e) $C_6H_5N_2^+Cl^-$,dil.HCl → X = $N_2C_6H_5$, 70%
(c) Br_2,CF_3CO_2H → X = Br 81%

Alkylation of 1,6-diazaphenalene using methyl iodide or benzyl bromide under conditions similar to those used for the alkylation of imidazole failed. However the reaction was successful using the lithium stabilised anion in tetrahydro-furan/hexamethylphosphoramide. Acylation of 1,6-diazaphenal-enes failed. However one characterisable amide (238) has been obtained (K. Avasthi and J.M. Cook, J. heterocyclic Chem., 1982, 19, 1415).

(238)   (239)

(j) *Other ring systems*

1,6-Diazabiphenylene (239) is obtained (38%) by flash vacuum photolysis of 2,5,9,10-tetraazaphenanthrene. This is the first example of an unsymmetrical diazabiphenylene (J.A.H. MacBride, P.M. Wright and B.J. Wakefield, Tetrahedron Letters, 1981, 22, 4545). The photo-dimerisation of *N*-methyl-2-pyridone has been studied in water and in non-aqueous solvents and the four isomers (240 - 243) have been obtained in the yields shown (Y. Nakamura, T. Kato and Y. Morita, J. chem. Soc. Perkin I, 1982, 1187).

trans,anti,51%   trans,syn,0.6%   cis,anti,11.2%   cis,syn,6.8%

(240)   (241)   (242)   (243)

Quindolines (245) have been prepared by the cyclocondensation of diketones of type (244) with α,β-unsaturated ketones. Dehydrogenation of the products gave the fully aromatised derivatives (V.P. Sevodin *et al.*, Khim. Getertsikl. Soedin, 1984, 1667).

(244)   (245)

R=Me,Ph,4-ClC$_6$H$_4$;   R^1=H,Me; R^2=Me,Ph,4-ClC$_6$H$_4$;   R^3=Me,Et,Ph.

Treating benzazepinones (246) with hot phosphorus oxychloride
gives not only the benzazepinopyrroloquinolines (247) but
also, via an unusual rearrangement, the quinopyrroloquino-
lines (248) (M.B. Stringer *et al*., J. chem.Soc. Perkin I,
1984, 2529).

R=Me; R^1=H,Me,MeO

(246)   (247)   (248)

Chapter 40

COMPOUNDS CONTAINING A SIX-MEMBERED RING HAVING TWO
HETERO-ATOMS FROM GROUP VIB OF THE PERIODIC TABLE:
DIOXANES, OXATHIANES AND DITHIANES

MALCOLM SAINSBURY

1.   Dioxanes and related compounds

The organisation and the nomenclature used in this chapter
follow the pattern established in the main work, C.C.C. 2nd
edn., Volume IVH, pp375-426.

(a) 1,2-Dioxanes, 1,2-dioxins and related compounds

(i) 1,2-Dioxanes
   1,2-Dioxanes are usually formed by the photochemical
addition of oxygen to alkenes;   antimony (V) chloride is
recommended as a catalyst (R.K. Haynes, M.K.S. Probert and
I.D. Wilmot, Austral.J.Chem., 1978, *31*, 1737).   Other
methods include the cyclisation of halogenated peroxides (1)
with silver (I) ions (N.A. Porter, J.C. Mitchell and P.M.
Gross, Tetrahedron Letters, 1983, *24*, 543), and the addition
of di-(tributyltin)peroxide to suphonate esters (e.g.2)
(R.G. Salomon and M.F. Salomon, J.Amer.chem.Soc., 1977, *99*,
3501).

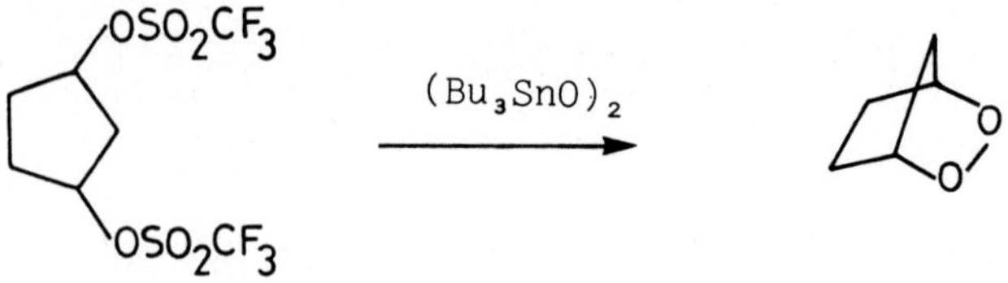

2

1,2-Dioxanes occasionally occur as natural products and as an example it has been found that sponges from the genus Chondrilla indigenous to the Great Barrier Reef in Australia give rise to the novel peroxyketal chondrillin (3), m.p.30°C (R.J.Wells, Tetrahedron Letters, 1976, 2637).

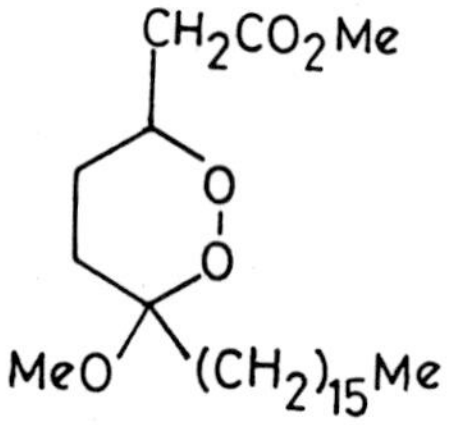

3

(ii) 1,2-Dioxins(1,2-dioxenes)

1,2-Dioxins can be obtained through the photosensitized oxygenation of cisoid 1,3-butadienes, (9E,11E) -methyl octadeca-9,11-dienoate (4), for example, affords the unsaturated cyclic peroxide (epidioxide) (5). This compound undergoes ready rearrangement in the presence of ferrous ion to the furanyl ester (6), probably by the mechanism shown below (E. Bascetta, F.D. Gunstone and C.E. Scrimgeour, J.chem.Soc.Perkin I, 1984, 2199).

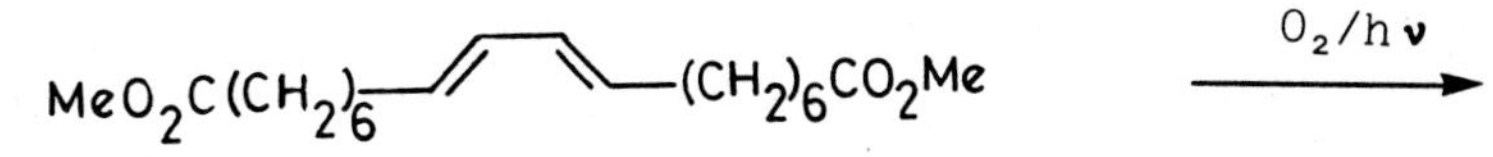

$MeO_2C(CH_2)_6$———$(CH_2)_6CO_2Me$ $\xrightarrow{O_2/h\nu}$

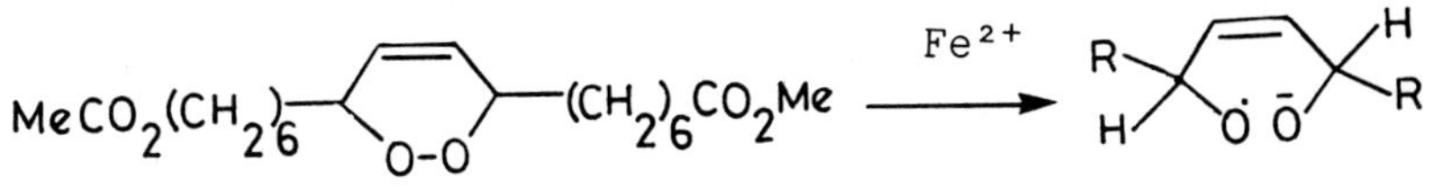

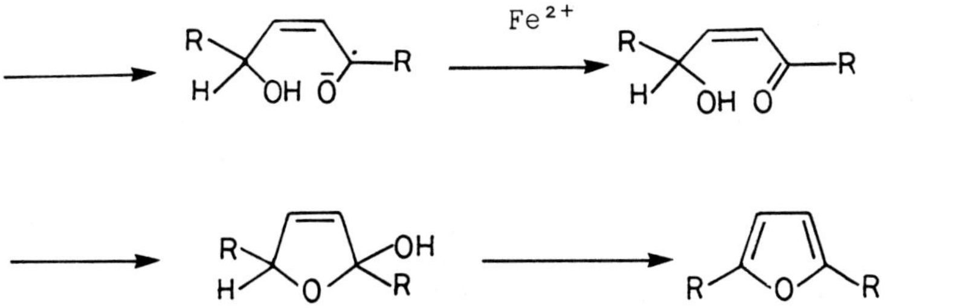

4

$MeCO_2(CH_2)_6$———$(CH_2)_6CO_2Me$ $\xrightarrow{Fe^{2+}}$

5

$\xrightarrow{Fe^{2+}}$

$R=(CH_2)_6CO_2Me$

2-Cyclohexene-1,4-endo-peroxide (7) (norascaridole) can be reduced to the cyclohexane analogue (8) by treatment with dipotassium azodicarboxylate. The product is a solid, m.p. 117-118°C, which deteriorates on standing at room temperature within a few days (W. Adam and H.J. Eggelte, Angew.Chem.internat.Edn.,1977, 713).

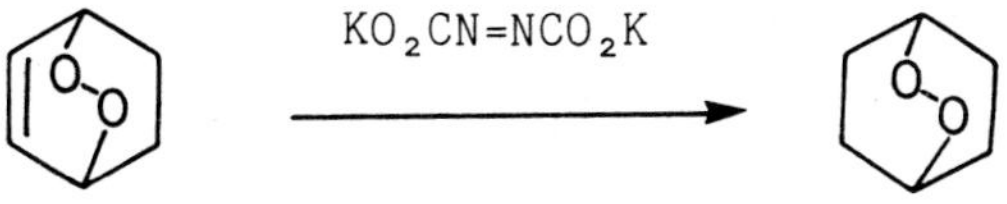

$\xrightarrow{KO_2CN=NCO_2K}$

7   8

Singlet oxygen addition to 5,6-dihydro-2-(2'-anthryl)-1,4-dioxin (9) affords the endo-peroxide (10); however, the peroxide undergoes quantitative rearrangement to the 1,2-dioextane (11) on treatment with silica in 1,2-xylene (A.P. Schaap, P.A. Burns and K.A. Zakilka, J.Amer.chem.Soc., 1977, *99*, 1270).

9   10

11

The dioxins (14) and (15) are prepared by the reaction of potassium dioxide upon the dibromides (or diiodides) (12) and (13) respectively (L.H. Dao et al., Can.J.Chem., 1977, *55*, 3791). The benzodioxin explodes violently when attempts are made to isolate it, but the naphthalide is quite stable and can be crystallised. The latter has m.p. 206-7°C, and was first made by F. Mayer, W. Schäfer and J. Rosenbach (Arch.Pharm., 1929, 584).

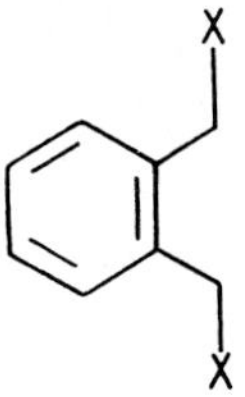

12

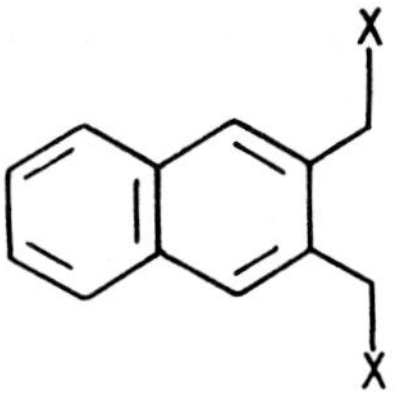

13

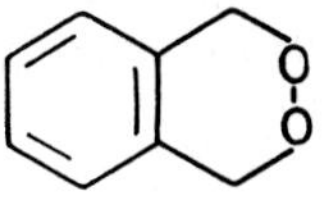

14

X=Br or I

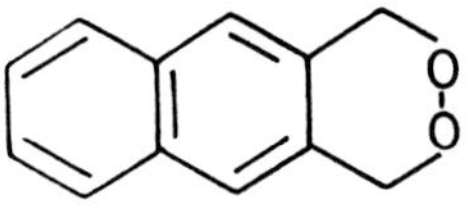

15

(b) 1,3-Dioxanes, 1,3-dioxins and allied compounds

(i) 1,3-Dioxanes

Synthesis.   Traditionally 1,3-dioxanes are prepared by reacting 1,3-glycols with aldehydes, or ketones, or their acetals in the presence of acid catalysts (G.M. Filippova, C.A., 1978, *89,* 43271u) or cation exchange resins (D.L. Rakhmankulov, ibid., 1980, *92,* 198406c). 5-Amido-1,3-dioxanes (17) may be synthesised, for example, by treating the diols (16) with methyl formate and acetone dimethyl-acetal in acetone containing hydrogen chloride (I.C. Nordin and J.A. Thomas, Tetrahedron Letters, 1984, *25,* 5723).

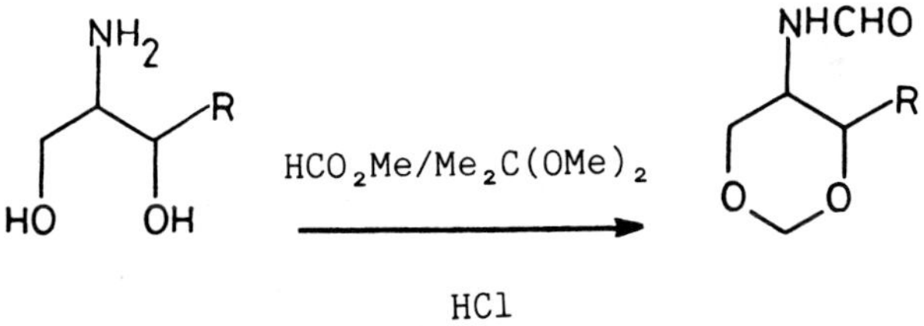

16

17

4,4-Dimethyl-1,3-dioxane (18) is obtained by reacting an aqueous solution of formaldehyde and isobutene with sulphuric acid and a transition metal (II) sulphate in a pressure vessel at 60°C (N. Yoshimura(Patent), C.A., 1977, *87*, 39497p). Similarly 4-methyl-5-nitro-1,3-dioxane is formed from the combination of 1-nitropropene and formaldehyde (Patent), C.A., 1977, *87*, 39386b).

Aryl methyl ketones and araldehydes react together in the presence of trimethylsilyl polyphosphate to give *meso*-2,4,6-triaryl-5-acyl-1,3-dioxanes (19) in good yields ($\approx$70%) under very mild conditions (T. Imamoto, H. Yokoyama and M. Yokoyama, Tetrahedron Letters, 1982, *23*, 1467).

$$(CH_3)_2C{=}CH_2 \ + \ HCHO \ \xrightarrow{\ H_2SO_4/M^{2+}\ }$$

18

$$ArCOMe \ + \ Ar'CHO \ \longrightarrow$$

19

Cyclocondensation of cyclohexanones (20) and formaldehyde in acetic acid containing acetic anhydride and phosphoric acid yields a mixture of tetraoxatricyclotetradecanes (21), dioxaspiroundecanones (22) and tetraoxatricyclotetradecanes (23) (F. Hirano and S. Wakabayashi, Bull.chem.Soc.(Japan), 1979,*52*, 779).

More control is observed in the base promoted reactions of formaldehyde with acyclic ketones (24; R and R'= alkyl, or aryl), which afford 5,5-disubstituted 1,3-dioxanes (25) (M. Delmas et al., Synthetic Comm., 1980, *10*, 517).

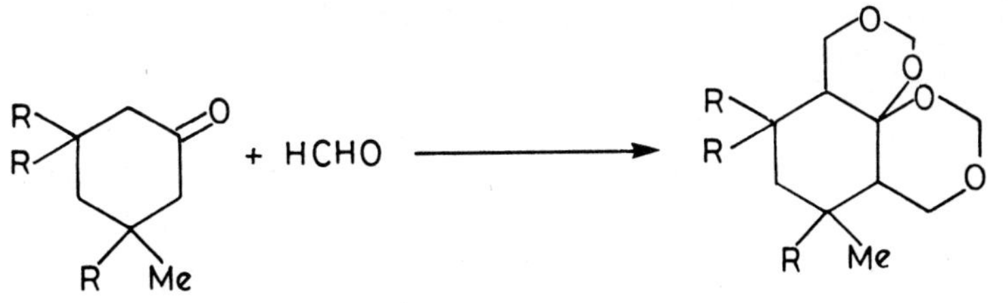

20 + HCHO ⟶ 21

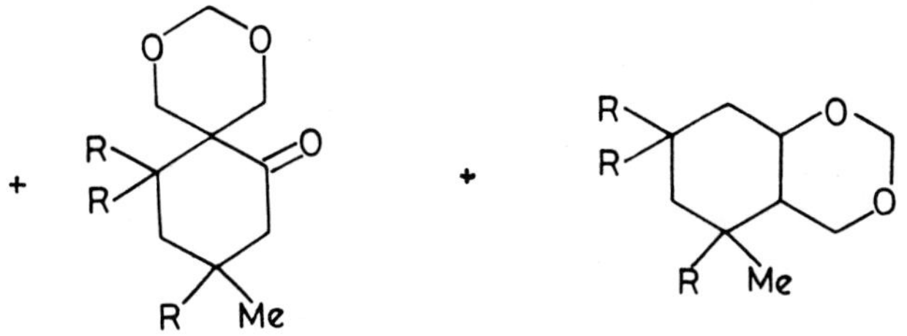

+ 22 (R=H, or Me) + 23

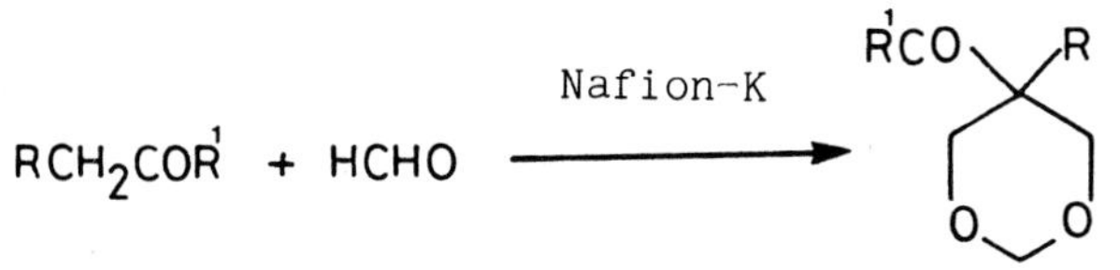

RCH_2COR^1 + HCHO $\xrightarrow{\text{Nafion-K}}$ 25

24

The spiro derivative (27) is formed by heating the tetraol (26) with formaldehyde at 145°C with sulphuric acid (A.C. Poshkus,(Patent) C.A., 1980, *93*, 26446v), whereas the isomer (29) is produced by reacting propan-1,3-diol (28) with methyl orthocarbonate in the presence of acids (T. Endo and M. Okawara, Synthesis, 1984, 837).

1,3-Dioxan-2-one, trimethylene carbonate, (30) is obtained through the reaction of oxetane, carbon dioxide and tetraphenylantimony(IV) iodide (A. Baba, H. Kashiwagi and H. Matsuda, Tetrahedron Letters, 1985, *26*, 1323).

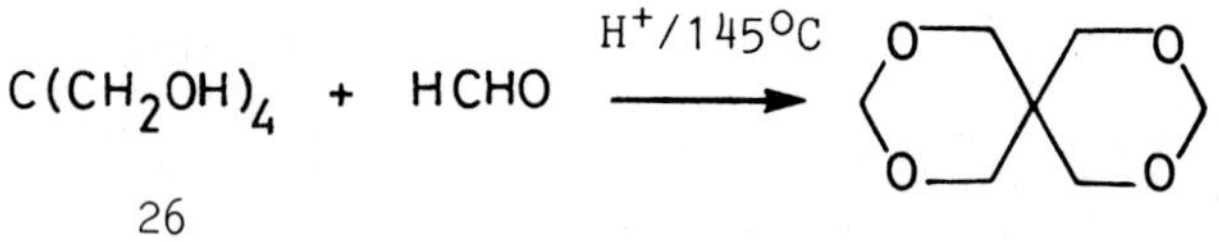

$$C(CH_2OH)_4 \; + \; HCHO \quad \xrightarrow{H^+/145^\circ C}$$

26

27

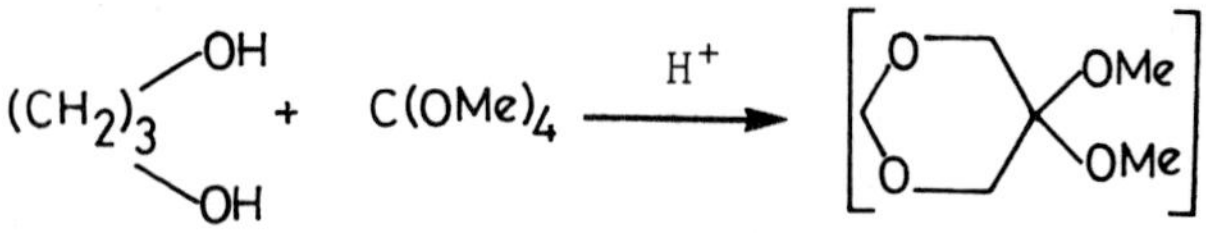

$$(CH_2)_3 \begin{smallmatrix} OH \\ \\ OH \end{smallmatrix} \; + \; C(OMe)_4 \quad \xrightarrow{H^+}$$

28

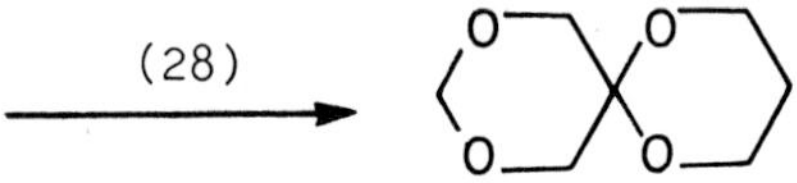

$$\xrightarrow{(28)}$$

29

$$\quad + \quad CO_2 \quad \longrightarrow$$

30

2-Substituted 5-oxo-1,3-dioxanes (32) are available through
the ozonolysis of the isopropylidene analogues (31)
(V. Imashev et al.,(Patent) C.A., 1980, *93*, 46689e), and
spirolactones (34) are obtained in 36-84% yield by treating
hydroxyacids (33) with ethoxyethene and zinc (II) chloride
(T. Fujita et al., Synthesis, 1979, 910).

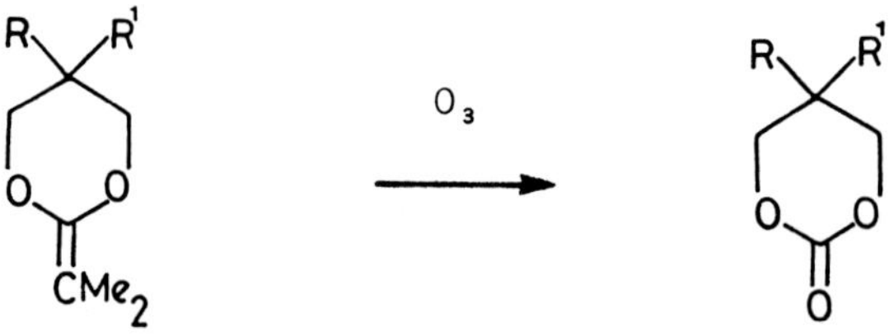

$$\xrightarrow{O_3}$$

31   (R,R^1=H, alkyl, or aryl) 32

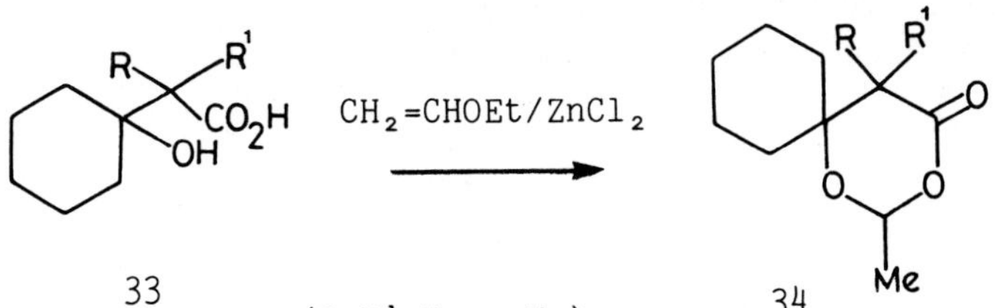

33

(R,R^1=H or Me)

34

Conformation. 1,3-Dioxanes normally adopt a chair conformation, but tub and twist representations occur under certain conditions (Z.I. Zelikman et al., Khim.Geterot. Soedin., 1978, 1172). In the case of 4,6-dioxo-1,3-dioxanes (35) the most stable conformation is a boat (P. Ayras and A. Partanen, Finn.chem.Letters. 1976, *415*, 110).

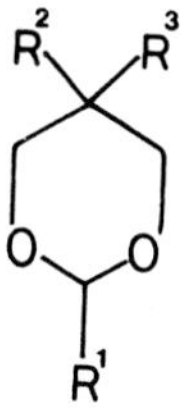

(R^1=R^2=H or Me; R^3=alkyl or Ph)

35

Ring-opening reactions. Reduction of 4-phenyl-1,3-dioxan with sodium/potassium alloy yields 2-phenylbutan-1,4-diol together with some 3-phenylpropanol and 3,4-diphenylhexan-1,6-diol (W.F. Bailey and E.A. Cioffi, Chem.Comm., 1981, 155).

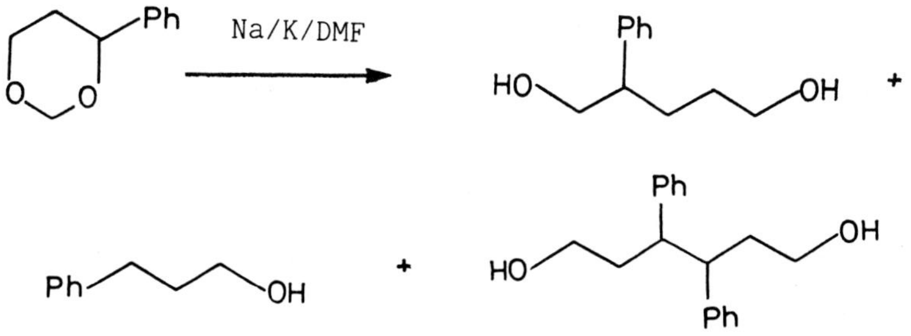

1,3-Dioxanes, which can be easily made in either enantiomeric form, are often used as chiral auxilliaries in synthesis. Thus, for example, acetals (36) react with organometallic reagents in the presence of titanium (IV) chloride to afford hydroxyethers (37) and (38). After separation, individual hydroxyethers may then be oxidised with pyridinium chlorochromate (PCC) and the residue of the auxilliary removed through a β-elimination process to afford the corresponding chiral alcohol. Enantiomeric excess is ≈70% (S.D. Lindell, J.D. Elliott and W.S. Johnson, Tetrahedron Letters, 1984, *25*, 3947).

When 1,3-dioxanes (39) are heated with p-toluenesulphonic acid or suphuric acid in an inert solvent they ring-open and then recyclise to pyranols (40) (N.A. Romanov et al., Zh.Prikl.Khim.(Leningrad), 1983, *56*, 333). In the presence of ethyl acetate, however, a mixture of products is produced including acetoxylated dienes (I.L. Rakhmankulov, E.A. Kantor and D.L. Rakhmankulov, ibid., 1980, *53*, 1367).

39

40

(R= alkyl, aryl or heteroaryl)

Related reactions occur when the substrates are heated with alcohols (M.G. Safarov, V.G. Safarov and S.R. Rafikov, Izv.Akad.Nauk.S.S.S.R. Ser.Khim., 1977, 1350) and with nitriles of isothiocyanates. In the last reactions the products are 4H-5,6-dihydro-1,3-oxazines (A.A. Gevorkyan, G.G. Tokmadzhyan and L.A. Saakyan, Arm.khim.Zh., 1977, *30*, 748).

Reactions of ring substituents. 5-Bromo-5-nitro-1,3-dioxane reacts with lithium enolates to afford alkylated products and structures such as (41) and (42) resulting from self condensation (V.V. Zorin et al., Zh.obsch.Khim., 1984, *54*, 1675).

41

42

Meldrum's acid. Meldrum's acid (isopropylidene malonate) (43) is a useful synthon and has attracted a good deal of interest (see H. McNab, Chem.Soc. Reviews, 1978, *7*, 345).

It may be alkylated reductively with aldehydes or ketones in the presence of borane/ dimethylamine complex. Thus condensation with benzaldehyde and reduction affords the benzyl derivative (44). (D.M. Hrubowchak and F.X. Smith, Tetrahedron Letters, 1983, *24*, 2951).

43   44

The initial condensation of Meldrum's acid with aldehydes to yield alkylidene derivatives is promoted by pyrrolidine (R. Chhabra et al., Austral.J.Chem., 1984, *37*, 1795), and these products react with Grignard reagents to give 2,2-dialkylated structures (M.L. Haslego and F.X. Smith, Synthetic Comm., 1980, *10*, 421; see also F.E. Ziegler, T. Guenther and R.V. Nelson, ibid., p.661). Copper catalyses this reaction (X. Huang, C-Chu and Q.-L.Wu, Tetrahedron Letters, 1982, *23*, 75).

The 2-benzoyloxy – Meldrum's acid (45) on heating to 460°C breaks down to a number of smaller molecules (R.F.C. Brown et al., Austral.J.Chem., 1976, *29*, 1705). Fragmentation also occurs when diazo Meldrum's acid (46) is heated: but photolysis in wet benzene affords the ketene (47), which may be trapped by the water present to give the unstable acid (48) (S.L. Kammula et al., J.org.Chem., 1977, *42*, 2931). Similar results have been reported by V.A. Nikolaev, N.N. Khimich and I.K. Korobitsyna (Khim.Geterot.Soedin, 1985, 321).

Methylene Meldrum's acid (50) is obtained from diisopropylidene methylene dimalonate (49) by reaction with formaldehyde (J.F. Buzinkai, D.M. Hrubowchak and F.X. Smith, Tetrahedron Letters, 1985, *26*, 3195).

(ii) 1,3-Dioxins and 1,3-dioxinones

Preparation.   When the methylene dioxane (51) is treated
with s-butyllithium and then with electrophiles, such as
aldenydes, ketones, alkyl bromides or epoxides, mixtures of
1,3-dioxins and 6-substituted methylene dioxanes are formed
(A.P. Kozikowski and K. Isobe, Tetrahedron Letters, 1979,
833).

A more selective preparation of 1,3-dioxins requires the
reaction of hexafluoroacetone and vinyl-aldehydes or ketones
(A.V. Fokin, A.F. Kolomiets and A.A. Krolevets, Izv.
Acad.Nauk.S.S.S.R.Ser Khim., 1978, 976).

$$CF_3COCF_3 \quad + \quad CH_2{=}CHC(R)O \quad \longrightarrow$$

(R=H or alkyl)

A general method for the synthesis of 1,3-dioxin-4-ones has been reported by M. Sato et al., (Chem.pharm.Bull.(Japan), 1983, *31*, 1896). In this approach β-keto acids are treated with a mixture of acetone, acetic anhydride and concentrated sulphuric acid or with isobutenyl acetate and sulphuric acid to give the dioxinones in good yield. (For related work see Sato et al., Heterocycles, 1984, *22*, 2563.)

$$\overset{2}{R}COCH(\overset{1}{R})CO_2H \quad \xrightarrow{MeCOMe/Ac_2O/H^+}$$

6-Phenyl-1,3-dioxin-4-ones are prepared by treating the furandione (52) with ketones (Yu S. Andreichikov, L.F. Genin and V.L. Genin, Khim.Geterot.Soedin., 1979, 1280). This is a version of an earlier synthesis, see G. Kollenz et al., (Z. Naturforsch-B, 1977, *52 B*, 701). Ketenes are presumed to be intermediates in the reactions.

$$52 \quad \xrightarrow[\Delta]{R^1COR^2}$$

When heated 1,3-dioxin-4-ones themselves decompose affording
acyl ketenes which can be captured with isocyanates
(G. Jäger and J. Wenzelburger, Annalen, 1976, 1689), imines
(M. Sato et al., Chem.pharm.Bull., 1983, *31*, 1902) and
enamines affording a variety of heterocycles(idem.ibid.,
p.4300).

(iii) Benzo-1,3-dioxins

Benzo-1,3-dioxins (e.g. 53) are well established compounds
which can be prepared by condensing 2-hydroxybenzyl alcohols
with aldehydes. The acetal proton can be removed by
treatment with triphenylmethyl perchlorate leading to the
salts (54), which in turn react with alcohols to yield
2-alkyloxy derivatives (55) (G.N. Dorofeenko and
S.M. Etmetchenko, Zh.org.Khim., 1976, *12*, 2228).

Benzo-1,3-dioxins unsubstituted at C-2 can be synthesised from phenols by reaction with two equivalents of formaldehyde in the presence of ion exchange resins such as Amberlite IR 120 and Nafion S01 (A. Denis et al., J.heterocyclic Chem., 1984, 21, 517).

Benzo-1,3-dioxin-4-ones (56) are formed by the reaction of dibromomethane with 2-hydroxybenzoic acids (L. Bonsignore, A.M. Fadd and G. Loy, C.A., 1979, 91, 193238u), and benzene-diazonium carboxylate combines with vinylene carbonate at 35-45°C to afford spiro structures (57) (J.M. Rao and S. Mallikarjuna, Tetrahedron Letters, 1979, 283).

Salicyclic carbonate (benzo-1,3-oxin-2,4-dione) (58) may be
alkylated at C-2 by a reaction with a β-keto-esters
(L.A. Mitscher et al., Heterocycles, 1978, *11*, 489), whereas
benzo-1,3-oxin-4-ones (59) are cleaved by treatment with
phenylmagnesium bromide (S. Melis and F. Sotgiu, C.A.,1979,
91, 193237t).

(c) 1,4-Dioxanes, 1,4-dioxins and 2,3-dihydro-1,4-dioxins

(i) 1,4-Dioxanes

Preparation. 2-Methoxy-1,4-dioxanes (62) are synthesised
by Kolbe-type anodic ring-expansion reactions of the acetals
(60). It is assumed that the intermediate cations (61) are
formed, after decarboxylation, and these rearrange prior to
capture of methanol – the electrolysis solvent
(D. Lelandais, C. Bacquet and J. Einhorn, Chem.Comm., 1978,
194; Tetrahedron, 1981, *37*, 3131).

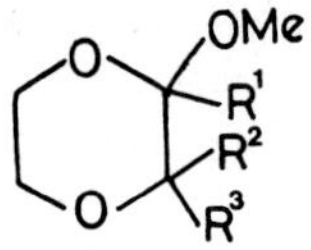

62

Treatment of diglycollic aldehyde (63) with acetic anhydride and concentrated sulphuric acid gives *cis*-2,6-diacetoxy-1,4-dioxane (64). However, acetic anhydride and pyridine yield the *trans*- isomer and the "dimer" (65) (A.F.J. Lopez, A.V. Espinosa and B.F. Zorrilla, An.Quim., 1977, *73*, 725). Similarly methoxydiglycollic aldehyde and excess methanol in the presence of concentrated sulphuric acid give a mixture of isomeric 2,3,5-trimethoxy-1,4-dioxanes (idem.ibid., p.721).

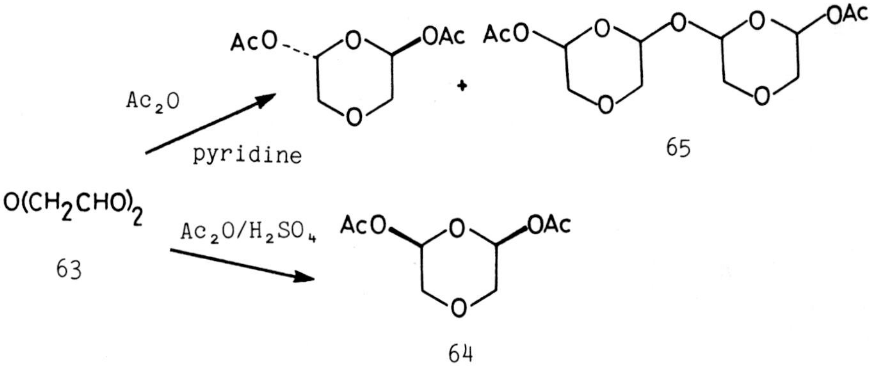

Glyoxal reacts with either alkyl or aryl chloroformates to afford tetrasubstituted 1,4-dioxanes (66) (K.F. Podraza, J.heterocyclic Chem., 1984, *21*, 1197).

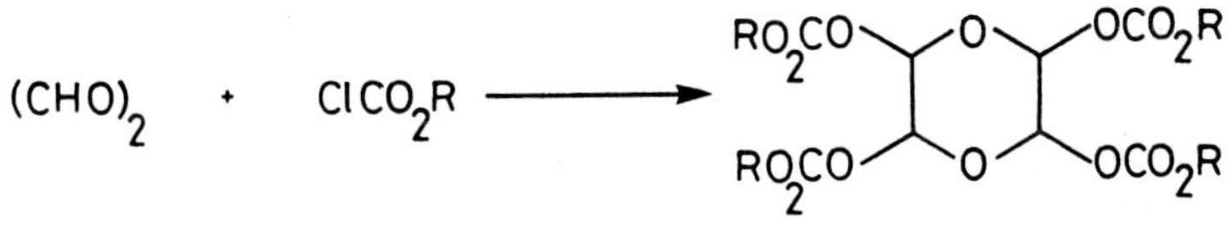

66

1,4-Dioxan-2-ones (68) undergo photochemical deoxygenation with trichlorosilane using *t*-butylperoxide as an initiator. The starting compounds may be obtained in two steps by treating diols with ethyl diazoacetate and rhodium (II) pivalate, followed by acid catalysed cyclisation of the intermediate esters (67) (N.G.C. Hosten, D. Tavernier and M.J.O. Anteunis, Bull.Soc.chim.Belg., 1985, *94*, 183).

67

68

Halogenated structures are popular targets for synthesis and 2-iodomethyl-1,4-dioxane is obtained by the reaction of allyl β-hydroxyethyl ether with potassium iodide and mercury (II) acetate (M.S. Boehringer, (Patent) C.A., 1977, *87*, 135350f). Trifluoro(trifluoromethyl)oxirane and perfluoro-propene, in di-β-methoxyethyl ether and hexamethylphosphoric triamide react together to yield the perfluoro-1,4-dioxanes (69) and (70) (T. Martini, Tetrahedron Letters, 1976, 1857). This procedure seems more efficient than direct fluorination of the substrates (see below).

69 R = CF(CF$_3$) COF

70 R = CF(CF$_3$) CF$_2$OCF(CF$_3$)COF

Reactions.   Electrochemical fluorination of 1,4-dioxane gives a poor yield of perfluoro-1,4-dioxane (71) plus fluorinated ring fragmentation products (V.V. Berenblit et al., C.A., 1980, *93*, 71655y).   However, fluorine and oxygen afford perfluoro-1,4-dioxane, pentafluoro-1,4-dioxane (72) and the ether (73) (J.L. Adcock, J.fluorine Chem., 1980, *16*, 297).
Pentafluorodioxane is also available through the action of colbalt (III) fluoride on 1,4-dioxane; it may be deprotonated by methyl lithium and the resultant anion then reacts normally with carbonyl compounds to afford a variety of products (P. Dodman and J.C. Tatlow, ibid., 1976, *8*, 263).

71

72

73

On photolysis in the liquid phase 1,4-dioxane gives a complex mixture of products, the nature of which indicate that the initial reaction involves rupture of a carbon-oxygen bond (J.J. Houser and B.A. Sibbio, J.org.Chem., 1977, *42*, 2145).   This is in conflict with earlier results (C.C.C. 2nd edn.Vol.IVH,p.383) which argue that carbon- hydrogen bond scission is the first step.

(ii) 1,4-Dioxins and 2,3-dihydro-1,4-dioxins
Oxidation of 2,3,5,6-tetraphenyl-1,4-dioxin with *m*-chloro-perbenzoic acid (MCPBA) leads to the acetal derivative (74)

of benzil (L. Lopez, V. Calo and M. Fiorentino, J.chem.Soc.Perkin I, 1985, 457). A similar result is obtained when this substrate is irradiated with ultraviolet light in an oxygenated benzene solution (M.V. George, Ch.V. Kumar and I.C. Scaiano, J.phys.Chem., 1979, *83*, 2452), but in this case the product is benzil itself. Ozonolysis of the parent compound also affords benzil.

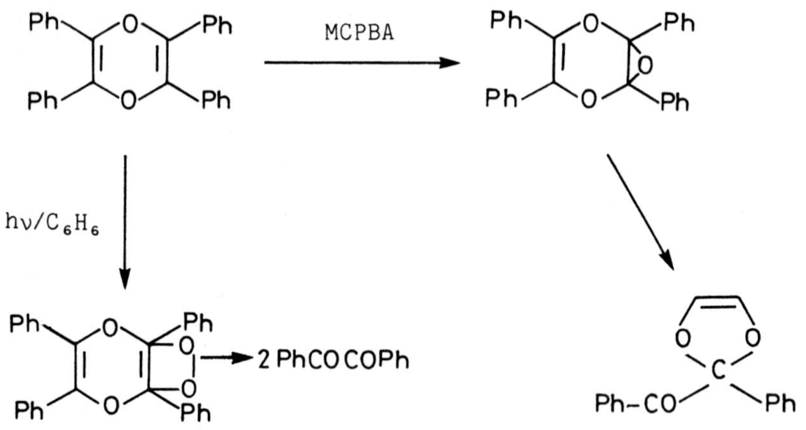

Dichlorocarbene adds to 1,4-dioxin in sequential fashion first to give the 1:1 adduct (75) and then the 2:1 product (76) (W. Schroth and W. Kaufmann, Z.Chem., 1978, *18*, 15). The adducts on treatment with base in alcoholic solution undergo ring-expansion to dioxepines (idem.ibid., 1977, *17*, 331).

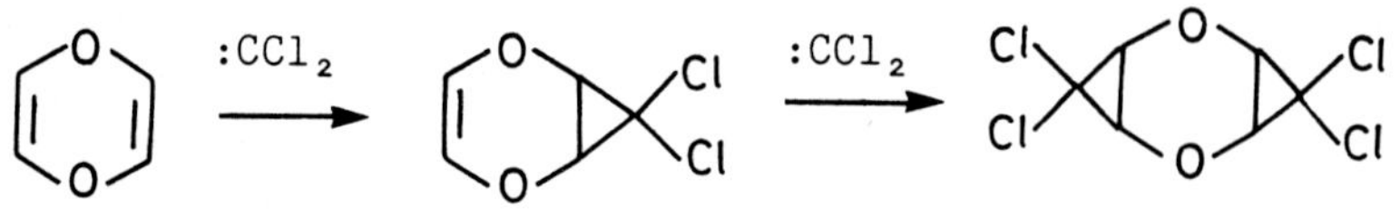

75

76

2,3-Dihydro-1,4-dioxin behaves similarly and with ethoxy-carbonyl carbene reacts to yield a mixture of the dioxabi-cycloheptane (77) and the dioxenylacetate (78) (S. Shatzmiller and R. Neidlein, Ann., 1977, 910).

2,3-Dihydro-1,4-dioxins are conveniently made by treating 2-methoxydioxanes (79) with phosphoric acid in pyridine (C. Bacquet, J. Einhorn and D. Lelandais, J.heterocyclic Chem., 1980, *17*, 831), whereas a photochemically induced addition of the diketone (80) and 1,2-difluoroethene gives a mixture of the dihydrodioxin (81) and the oxetane (82) (M.G. Barlow, B. Coles and R.N. Hazeldine, J.chem.Soc.Perkin I, 1980, 2523).

2,3-Dihydro-1,4-dioxin, prepared by the cyclodehydration of di-β-hydroxyethyl ether, undergoes a Vilsmeier reaction to afford the formyl derivative (83), and adds hydrogen chloride to yield 2-chloro-1,4-dioxane (84) (N.V. Kuzetsov and I.I. Krasavtsev, Ukr.khim.Zh., 1976, *42*, 1063). When it is irradiated with ultraviolet light in the presence of benzene the tricycle (85) is produced in only moderate quantum yield (J. Mattay, H. Leismann and H.-D. Scharf, Ber., 1979, *112*, 577).

2,3-Dihydro-1,4-dioxin reacts with butyllithium to give the 5-lithio derivative which may then be used to synthesise 5-alkyldihydrodioxins by treatment with alkyl halides (R.W. Saylor and J.F. Sebastian, Synth.Comm., 1982, *12*, 579). Carbonyl compounds yield addition products (86) which, when protonated, ring-open to afford α,β-unsaturated aldehydes; these on reduction with lithium aluminium hydride give α-ketoalcohols (87), thus completing a two carbon homologation of the original carbonyl compounds (M. Fetizon, I. Hanna and J. Rens, Tetrahedron Letters, 1985, *26*, 3453; see also p.4925).

$$87$$

(iii) 1,4-Benzodioxin and its derivatives

Preparation.   1,2-Cyclohexadione reacts with 1,2-diols (88) in the presence of dichlorobis(benzonitrile)palladium (II) to give 1,4-benzodioxins substituted at positions 2 and 3, and the stereochemistry of the diol is retained in the product heterocycle (E. Minicione, A. Sirna and D. Covini, J.org.Chem., 1981, *46*, 1010).

$$88$$

Reactions.   Substituents such as hydroxyl or formyl groups attached to the heterocycle of 1,4-benzodioxin undergo the usual reactions expected of such functions (see N. Petragnani, T. Brocksom and A. Moro, Farmaco.Ed.Sci., 1977, *32*, 512;   W. Adam et al., Ber., 1983, *116*, 1686), whereas the benzenoid ring undergoes electrophilic substitution with the appropriate reagents (see V. Danksas et al., Khim.Geterot.Soedin., 1977, 467).

Structures of biological interest.   A number of natural products contain a 1,4-benzodioxin unit including those classed as lignans.   Silybin (89) from *Silybum marianum* Gaertn.(C.C.C.2nd edn.,Vol.IVH,p.387) has been synthesised (L. Merlini et al., Chem.Comm., 1979, 695).   Several antibiotic structures bearing the same heterocyclic unit are known (see L. Foley et al., J.Antibiot., 1979, *32*, 418), and considerable interest resides in the synthetic compound idazoxan (90) which is an adrenoceptor blocking agent (B. Gadie et al., Brit.J.Pharmacol., 1984, *83*, 707).

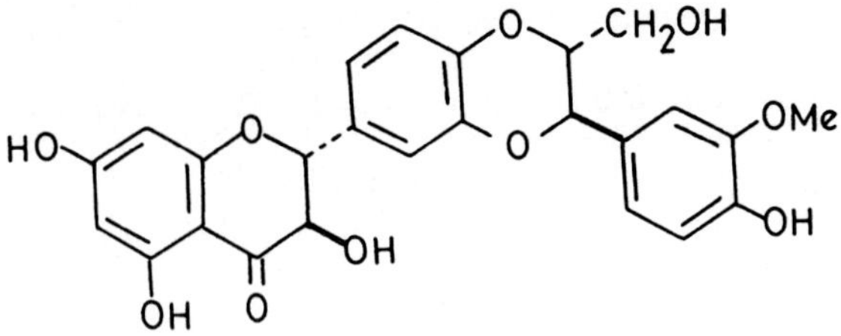

89

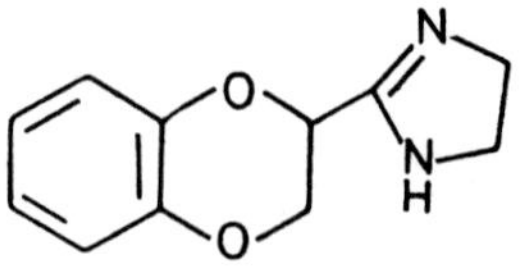

90

Concern over the effects of chlorinated dibenzodioxins as environmental contaminants still exists and a number of these structures have been synthesised by the cyclisation of polychlorobenzene derivatives as standards for analytical purposes (A.P. Gray et al., Trace. Subst. environ. Health, 1975, *9*, 255).

Ultraviolet irradiation of the hexachloro derivative (91) in benzene/hexane solution gives rise to the 1,2,3,6,8-, 1,2,3,7,8- and 1,2,3,6,7-pentachloro analogues (H.R. Buser, Chemosphere, 1979, *8*, 251).

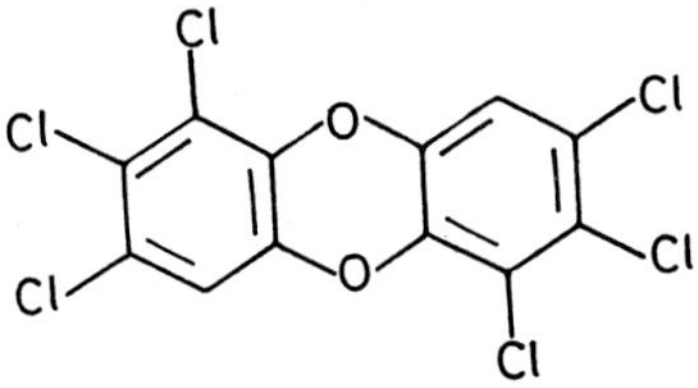

91

2.   Dithianes, dithiins and their benzo derivatives

(a) 1,2-Dithiins and benzo-1,2-dithiins

1,2-Dihydrodithiins bearing alkenyl substituents can be
synthesised in several ways.   Thus when a mixture of
myrcene (92) and sulphur is irradiated for several days the
pentenyl-3,6-dihydrodithin (93) is obtained together with
the thiophene (94) (T.L. Peppard and J.A. Elvidge,
Chem.Ind., 1979, 552).   Acrolein (95) hydrogen sulphide,
and ethyl orthoformate when heated together in ethanol
containing zinc chloride yields a mixture of the 1,2- and
1,3- dihydrodithiins (96) and (97) (P. Beslin, J.hetero-
cyclic Chem., 1983, *20*, 1753).

Thioketones (98) on flash thermolysis undergo a retro Diels
Alder reaction to give alkenes (99) which dimerise and
rearrange to give 1,2-dithiins (100) (Beslin et al.,
Tetrahedron, 1981, *37*, 3839).

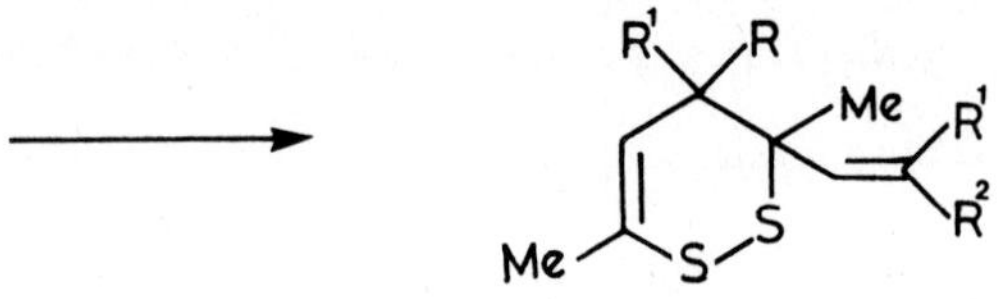

100

Methylene sulphone undergoes a cycloaddition reaction with
the dithiocarbamate (101) to produce the dihydrodithiin
dioxide (102) (J.C. Meslin, J.P. Pradere and H. Quinion,
Bull.Soc.Chim.Fr., 1976, 7-8, 1195).

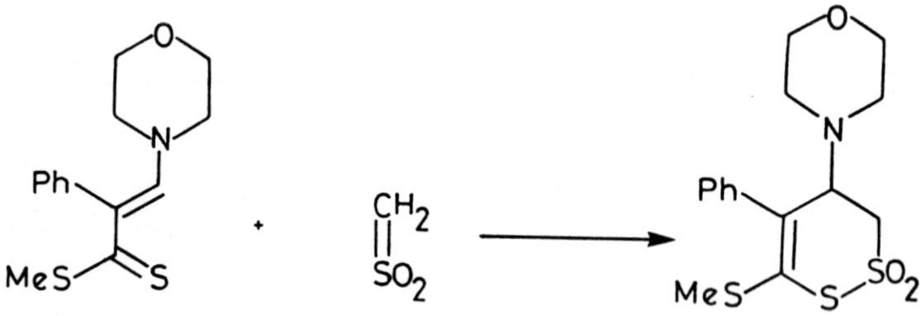

101   102

Very little work has been reported on the reactions of
1,2-dithiins: however, the benzodithiin (103) on treatment
with sodium methoxide ring contracts to the benzo(c)thio-
phene derivative (105).   In this reaction the thiol (104)
is an isolatable intermediate, and if it is heated in
methanol solution alone it affords the dihydrobithiophene
(106) which on heating at higher temperatures disproportion-
ates into the thiophene (105) and the reduced analogue (107)
(G. Cigarella, A. Nuvole and M.M. Curzu, Gazz., 1981, 111,
333).

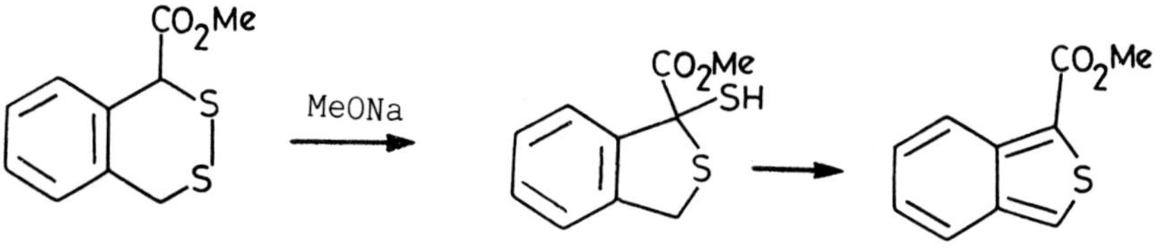

103   104   105

(b) 1,3-Dithianes and 1,3-dithiins

(i) 1,3-Dithianes

The use of 1,3-dithanes as masked carbonyl equivalents dominates most other aspects of their chemistry and much work has been undertaken to develop ways of functionalising the 2-position. Thus the corresponding anions formed by deprotonation with lithium alkyls have been shown to react with various trimethyl metal chlorides (Me$_3$MCl) to yield derivatives (108) which may then be acylated with acid chlorides (P. Jutzi and O. Lorey, Phosphorus Sulphur, 1979, 7, 203).

(M=Se, Ge or Sn)

The reaction between 2-methoxycarbonyl-1,3-dithiane and an equimolar mixture of pivalyl chloride and an aldehyde, followed by dealkylative decarboxylation of the resulting pivalate ester (109), gives ketene thioacetals (J.L. Belletire, D.R. Walley and S.L. Fremont, Tetrahedron Letters, 1984, 25, 5729).

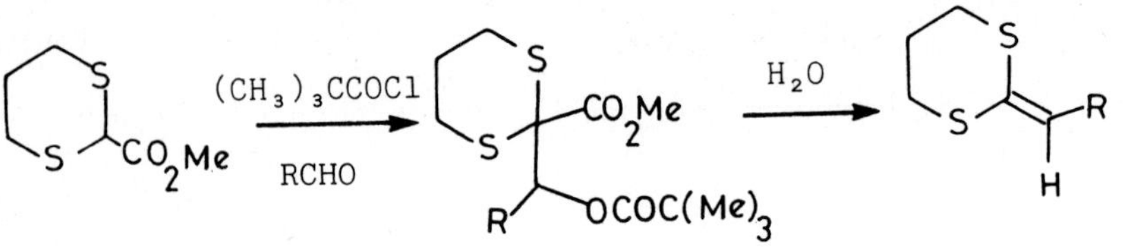

109

The anions (R^1=H, Me, Me$_3$Si) also react with cycloalkenones (110) to give the corresponding alcohols (111) C.A. Brown and A. Yamaichi, Chem.Comm., 1979, 100). However, in the case of the alkylation of vinyl species (e.g. R^1= CH=CHPh) both the hardness of the alkylating agent and the nature of the leaving group of the alkylating agent determine whether the reaction occurs at the α- or γ-position . Phenylmethyl iodide, for example, affords mainly the γ-product (112) whereas trimethylsilyl chloride yields the α-derivative (113) exclusively (W.S. Murphy and S. Wattanasin, J.chem. Soc.Perkin I, 1980, 2678).

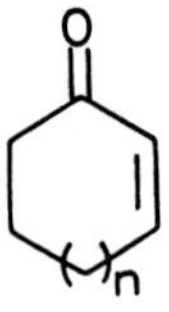

110 (n=0,1,2)

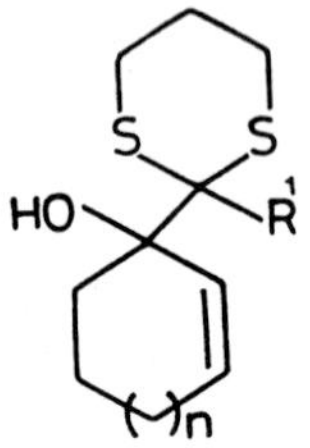

111

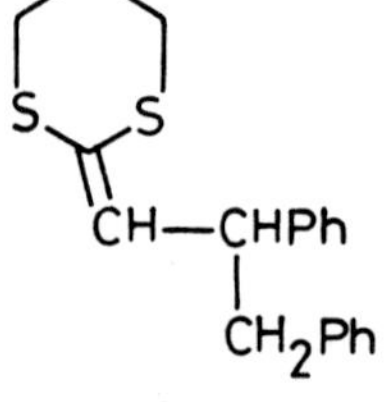

112

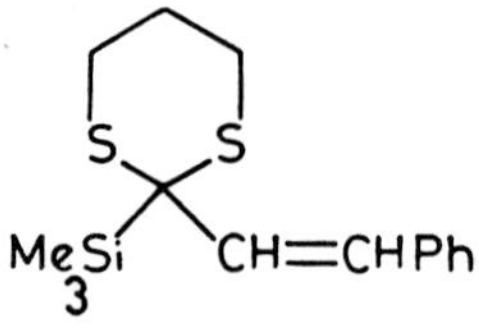

113

The anion of 2-trimethylsilyl-1,3-dithiane (114) combines
with nitriles to yield adducts (115) which rearrange and may
then be protonated to give primary amino thioacetals (116)
(P. Page, M.B. Vanniel and P.H. Williams, Chem.Comm., 1985,
742).

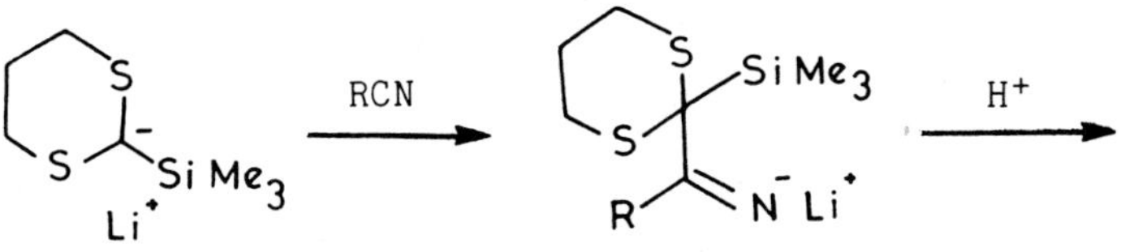

114   115

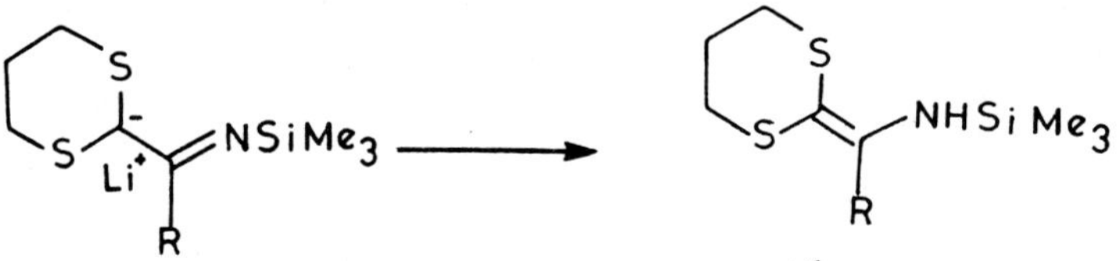

116

When treated with carbon disulphide 1,3-dithianyl lithium
gives the dilithium salt (117) which can be dialkylated with
alkyl halides (D.M. Baird and R.D. Bereman, J.org.Chem.,
1981, 46, 458). The anion also reacts with tropylium
fluoroborate to give the heptafulvene derivative (119),
after deprotonation of the intermediate (118) (K.M. Rapp and
J. Daub, Tetrahedron Letters, 1976, 2011).

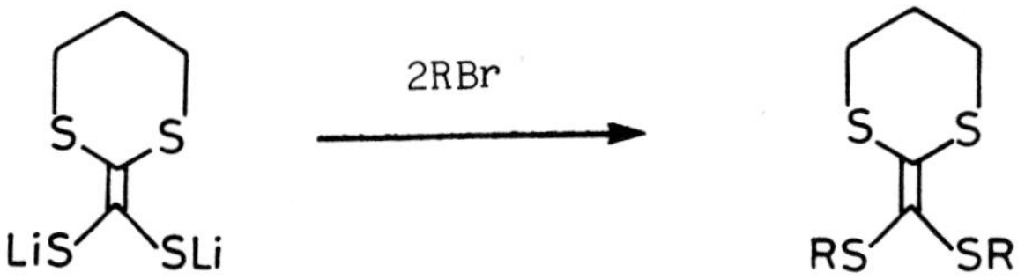

117

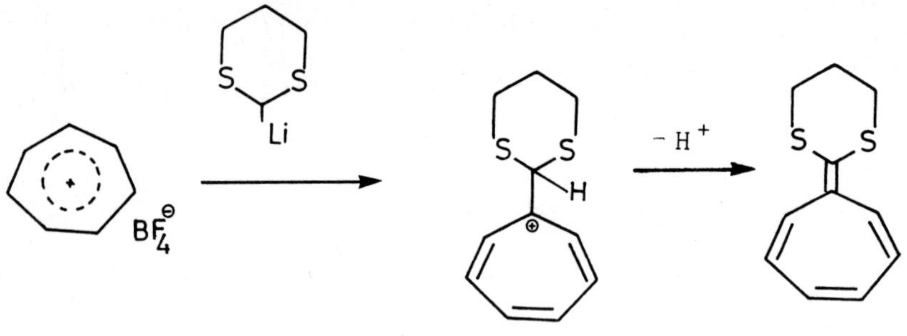

118   119

2-Cyano-1,3-dithianes can be converted into the lithium salts and these react with alkyl halides in the usual way. The parent structures are obtained from 1,3-dithianyl lithium by treatment with triphenylmethyl isonitrile (H.N. Khatri and H.M. Walborsky, J.org.Chem., 1978, *43*, 734).

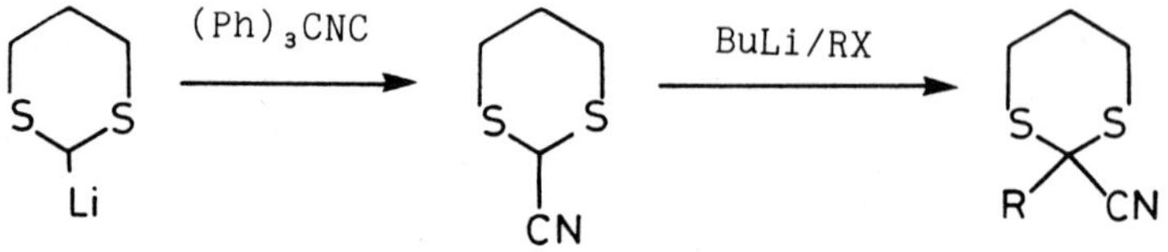

2,2-Disubstituted 1,3-dithianes are deprotonated by strong base at position 4, and then undergo Wittig-type rearrangements to the anions of 2,2-disubstituted tetrahydrothiophene-3-thiols, which may be trapped with alkyl halides to produce the 3-alkylthio derivatives (120) (H. Ikehira and S. Tanimoto, Bull.chem.Soc.(Japan), 1984, *57*, 2474).

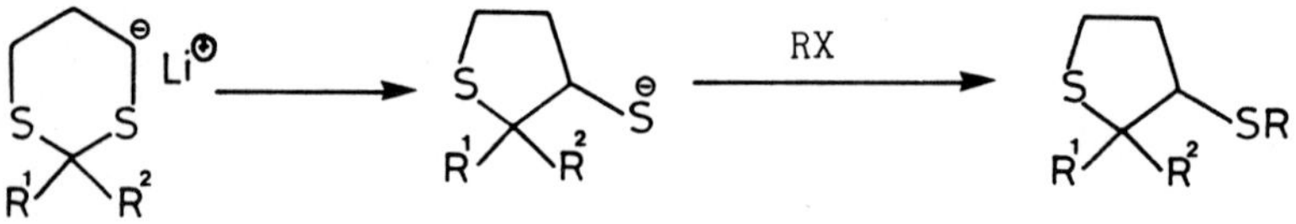

120

When the triphenylphosphonium salt (121) is reacted with propan-1,3-dithiol the Wittig reagent (122) is obtained thus providing a useful reagent for formylolefination (H.J. Cristan, H. Cristol and D. Bottaro, Synthesis, 1978, 826):

$$Ph_3\overset{\oplus}{P}CH=CH\overset{\oplus}{P}Ph_3 \cdot 2\overset{\ominus}{Br} \ + \ HSCH_2CH_2CH_2SH \ \longrightarrow$$

121

$$\overset{\oplus}{C}H_2\overset{\oplus}{P}Ph_3$$

122

A convenient and mild regeneration of carbonyl compounds from 1,3-dithiane-1-oxides can be effected by treating them with triethyloxonium tetrafluoroborate (I. Stahl et al., Angew Chem., 1979, *91*, 179). The oxides themselves are obtained from the corresponding 1,3-dithianes by oxidation with sodium periodate.

$$\xrightarrow{\text{NaIO}_4} \qquad \xrightarrow{(\text{Et})_3\text{OBF}_4}$$

$$\xrightarrow{\text{H}_2\text{O}} \text{RCHO}$$

2-Chloro-1,3-dithiane prepared by treatment of 1,3-dithiane with sulphuryl chloride reacts with Grignard reagents or with enolate anions to afford 2-substituted 1,3-dithianes (C.G. Kruse, A. Wijsman and A. Van der Gen, J.org.Chem., 1979, *44*, 1847). Organoaluminium reagents react similarly (G. Picotin and P. Miginiac, ibid., 1985, *50*, 1299). If the 1,3-dithianyl-2-ylium fluoroborate salts are prepared they react similarly with Grignard reagents, alkyl lithiums, or lithium alkyl cuprates (J. Klaveness and K. Undheim, Acta Chem.Scand. Ser B., 1983, *37*, 258; 687).

Salts of this type also react with allylsilanes leading to 2-alkenyl-1,3-dithianes (123) (C. Westerlund, Tetrahedron Letters, 1982, *23*, 4835).

2-Formyl-1,3-dithiane is readily alkylated by allyl halides (S.R. Wilson and J. Mathew, Synthesis, 1980, 625), and 2-chlorocarbonyl-1,3-dithianes can be obtained from the parent acids by treatment with oxalyl chloride (E.C. Taylor and J.L. LaMattina, J.org.Chem., 1978, *43*, 1200). Reaction of the acid chloride (124) with triethylamine leads to the ketene (125) (P. Schenone et al., J.heterocyclic Chem., 1976, *13*, 1105).

2-Alkylidine-1,3-dithianes undergo oxidative ring-expansion to 3-alkyl-1,4-dithiepan-2-ones (126) on treatment with lead (IV) acetate (K. Hiroi and S. Sato, Chem.Letters, 1979, 923).

126

1,3-Dithian-2-one is synthesised by reacting 1,3-dibromo-propane with potassium ethoxydithioformate in boiling ethanol (N.N. Godovikov et al., Izv.Akad.Nauk.S.S.S.R. Ser.Khim., 1976, 2384), whereas the corresponding 2-thione is prepared by reacting 1,3-dihydroxypropane with sodium trithiocarbonate (from CS_2 and Na_2S) (S. Kubota et al., Agric.biol.Chem., 1977, *41*, 1621).

5-Amino-1,3-dithianes (127) when treated with phosgene rearrange to 3-imino-7,7a-dihydro-1*H*,3*H*,5*H*-thiazolo[3,4-c]thiazoles (128) (J. Borgulya et al., Helv., 1984, *67*, 1827).

127   128

The general conclusions regarding the preference for the adoption of a chair conformation in simple 1,3-dithianes (C.C.C. 2nd edn., Vol.IVH, p.395) have been confirmed (A.T. McPhail, K.D. Onan and J. Koshimies, J.chem.Soc., Perkin II, 1976, 1004). The 1-oxides and 1,3-dioxides also favour this geometry and it is interesting that 1,3-di-thiane-1-oxide may be resolved into its enantiomers by forming the (+)- camphor adducts (B.F. Bryan et al., J.org.Chem., 1978, *43*, 90).

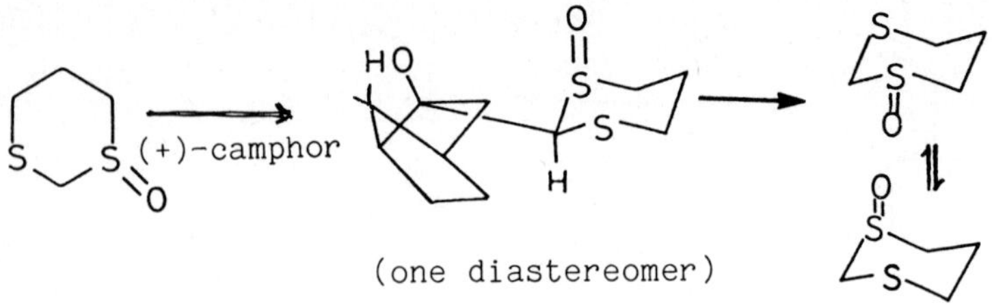

(one diastereomer)

(ii) 1,3-Dithiins

This class of structure is still relatively uncommon but 2-alkylimino-5-methyl-1,3-dithiins (130) are prepared in good yields by heating 2-dialkylamino-4-bromomethyl-4-methyl-1,3-dithiolanylium bromides(129) (K. Hiratani, T. Nakai and M. Okawara, Bull.chem.Soc.(Japan), 1976, *49*, 2339).

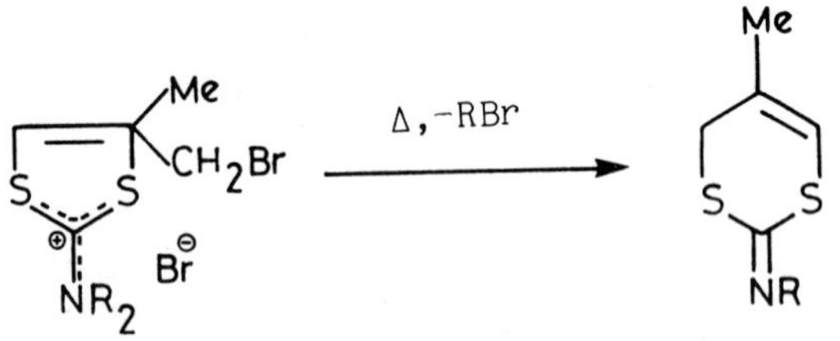

129   130

(c) 1,4-Dithiins, 2,3-dihydro-1,4-dithiin, 1,4-dithianes and
their benzo derivatives

(i) 1,4-Dithiins and their reduced forms

A general synthesis of 1,4-dithiins from diketosulphides requires the treatment of these substrates with either phosphorus (IV) sulphide or Lawesson's reagent and sodium hydrogencarbonate in boiling toluene (J. Nakayama et al., Heterocycles, 1984, *22*, 1527).

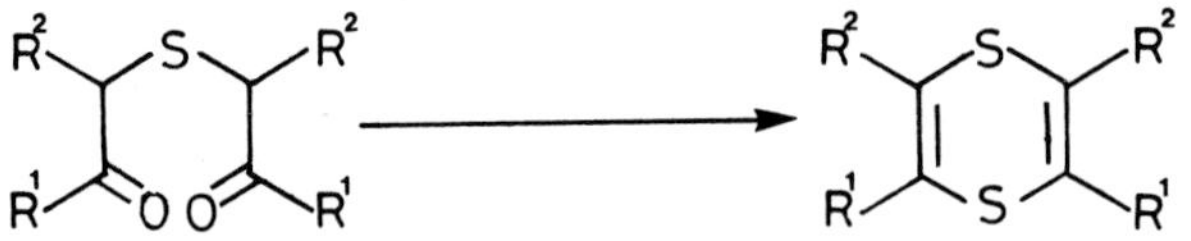

Dimethyltetrathiooxalate (131) and 2,3-dihydro-4-pyran yield the dihydrodithiin (132), but with phenylacetylene the dithiin (133) is obtained (K. Hartke, J. Quante and T. Kämpchen, Ann., 1980, 1482).

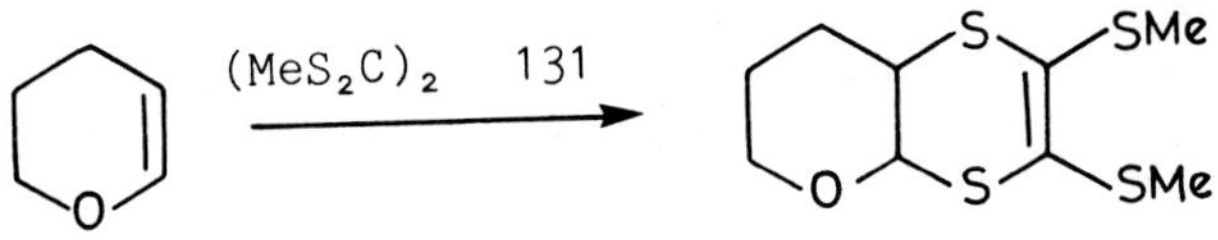

132

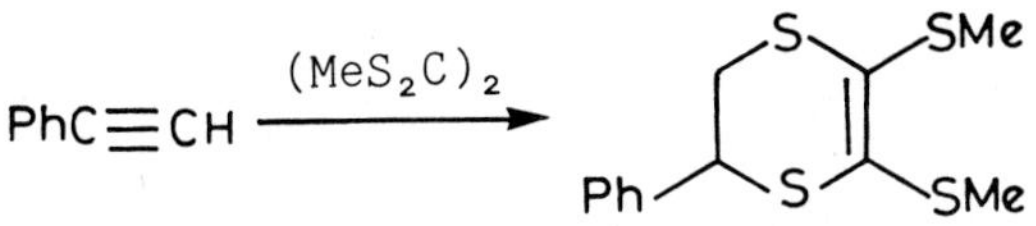

133

Dihydro-1,4-dithiins can be made by treating α-haloketones, or their acetals, with 1,2-ethandithiol (G. Giusti and G. Schembri, Compt.Rend., 1978, *287C*, 213; E.A. Ramazanov et al., Azerb.Khim.Zh., 1984, 52; C.A., 1985, *103*, 37429t), or by reacting alkenes with dimethyl dithiooxalate or dimethyl tetrathiooxalate. In this last reaction alkynes give the corresponding dithiins (K. Hartke et al. Angew Chem., 1978, *90*, 1016; Ann., 1980, 1482).

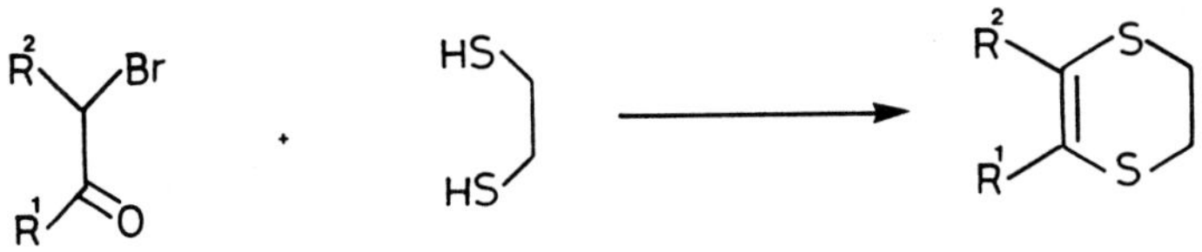

X=O or S

1,4-Dithiins react with *n*-butyllithium at -110°C in tetra-hydrofuran to give the 2-lithiated derivative which is suitable for elaboration through treatment with electro-philes at this temperature. However, as the temperature is raised to -60°C ring-scission occurs (M. Schoufs et al., Rec.Trav.chim., 1977, *96*, 259).

Ring-cleavage and ring-contraction occur when 2-nitro-dithiins (134) are *S*-methylated and treated with triethyl-amine. When oxidised to the *S*-oxides, these eliminate sulphur monoxide and the thiophenes (135) and (136) are formed (T.E. Young and A.R. Oyler, J.org.Chem., 1980, *45*, 933).

Benzyne reacts with 2,5-diphenyl-1,4-dithiin-1,1-dioxide to give 2-phenylbenzo[b]thiophene (137), whereas 1,1-di-(methoxycarbonyl)ethyne affords 2,3-di-(methoxycarbonyl)-5-phenylthiophene (138) (K. Kobayashi and K. Mutai, Tetrahedron Letters, 1978, 905).
The corresponding tetraoxide (139) with sodium azide yields the thiazine (140) (H.A. Levi et al. ibid., 1982, 23, 299).

(ii) Benzodithiins

Ring expansion of 1,4-benzoditholyl alcohols (141) through reaction with thionyl chloride and triethylamine gives 2-alkylidene 1,4-benzodithiins (142) (P. Blatcher et al., Tetrahedron Letters, 1978, 2349).
1,4,2-Benzodithiazines (143,R = H or Ph) eliminate R-CN and add dimethyl acetylenedicarboxylate to afford the benzodithiin (144) (J. Nakayama et al., Chem.Comm., 1982, 612).

141   142

[R^1,R^2 = H or alkyl; R^3 = H alkyl or phenyl;
R^3,R^4 = $(CH_2)_5$]

143   144

The sulphones (145) deprotonate on treatment with sodium
hydride and the conjugate anions react with carbon disulphide to form the salts (146). When treated with alkyl
chlorides these produce 1,4-benzodithiin-1,1-dioxides (147)
(W.D. Rudorf and D. Janietz, Synthesis, 1984, 854).

145   146

147

(X = H, 3-Cl, 4-Cl; R = H, Me, Ph, CO_2Me)

1,4-Benzodithiin-1,1,4,4-tetraoxide is an effective diene-ophile and undergoes cycloaddition reactions with, for example, anthracene, 1,3-pentadiene, cyclopentadiene and furan under mild conditions (J. Nakayama, Y. Nakamura and M. Hoshino, Heterocycles, 1985, *23*, 1119).

(iii) Thianthrenes
   Preparation.   When heated 1,2,3-benzothiadiazoles (148) fragment giving thianthrenes (149) (P.C. Montevecchi and A. Tundo, J.org.Chem., 1981, *46*, 4998).

(R = H, 6-Cl or 5-CO$_2$Me)

In such reactions there is evidence that the intermediate species is a benzothiirene (T. Woolridge and T.D. Roberts, Tetrahedron Letters, 1977, 2643).

A more useful approach to thianthrenes requires the reactions of 2-chlorobenzenethiols with triethylamine and hexamethylphosphoramide (I.W.J. Still and V.A. Sayeed, Synth.Comm., 1983, *13*, 1181).   2,7-Dinitrothianthrene is prepared by heating 2,2'-dichloro-5,5'dinitrodiphenyl-disulphide with sodium hydroxide in aqueous ethanol (V.N. Lisitsyn and S.A. Zaitsev, Zh.org.Khim., 1984, *20*, 202).   A route to dimethylthianthrenes employs reactions between 2-chlorotoluenes and sulphur monochloride in the presence of aluminium (III) chloride.   Unfortunately mixed isomers form.   (H.C. Lin (Patent), C.A., 1979, *90*, 186964d.)

Treatment of 1,2-dimethoxybenzene first with sulphur dichloride and then with tin (II) chloride gives the tetramethoxythianthrene (150) (T. Weiss and G. Klar, Ann., 1978, 785).   This compound has been subjected to X-ray crystallographic analysis (O.A. D'yachenko et al., C.A., 1978, *88*, 50750g).

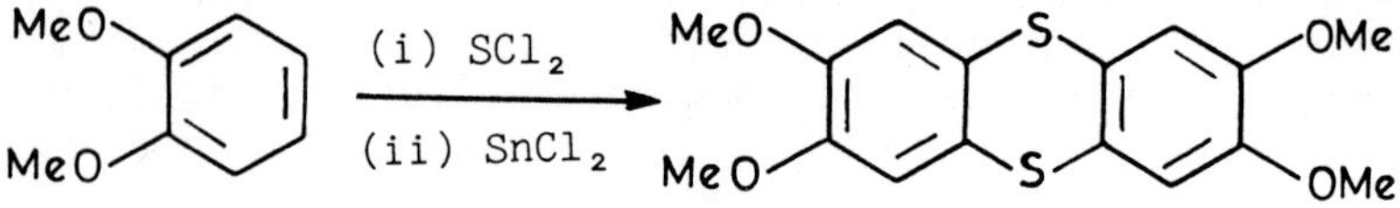

150

Thianthrene reacts with one molecular equivalent of diethyl diazomalonate to form the monosulphonium ylide (151). This structure is confirmed by single crystal X-ray studies: the angle of fold between the two aryl rings is 135.7° (A.L. Ternay Jr., et al., J.heterocyclic Chem., 1982, *19*, 833).

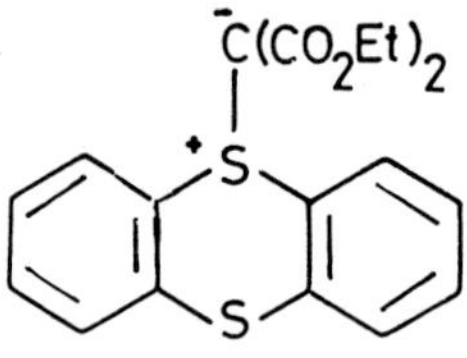

151

The radical cation of thianthrene (and also that of phenoxa-thiin) forms adducts linked through a sulphur atom with alkenes and alkynes (H.J. Shine et al., J.org.Chem., 1979, *44*, 915; 1981, *46*, 271).

(d) Telluranthrenes

Perfluorotelluranthrene (152) is synthesised by heating 3,4,5,6-tetrafluoro-1,2-diiodobenzene with tellurium, followed by treatment of the product oil with bromine and then with sodium sulphide (D.P. Rainville, R.A. Zingaro and E.A. Meyers, J.fluorine Chem., 1980, *16*, 245).

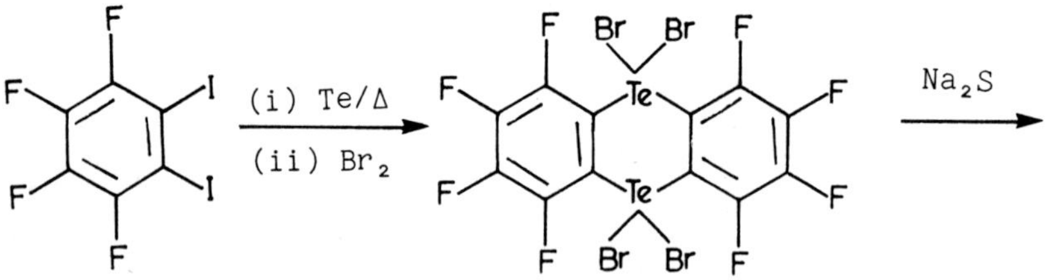

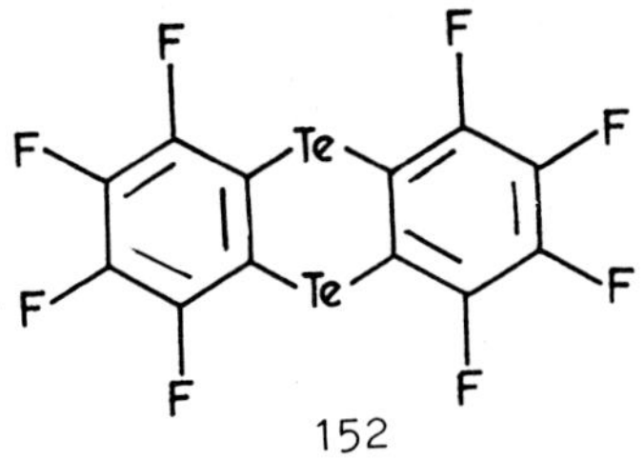

152

3. Oxathianes, oxathiins and related compounds

(a) 1,2-Oxathianes, 1,2-oxathiins and their benzo derivatives

(i) 1,2-Oxathianes

Cyclisation of 4-chloroalkylsulphuryl chlorides by heating in butanol gives 1,2-oxathiane dioxides (153) *via* the intermediacy of the corresponding butyl esters (A.M. Kurdynkov, A.P. Khardin and Yu A. Komyakov, C.A., 1980, *93*, 46545e).

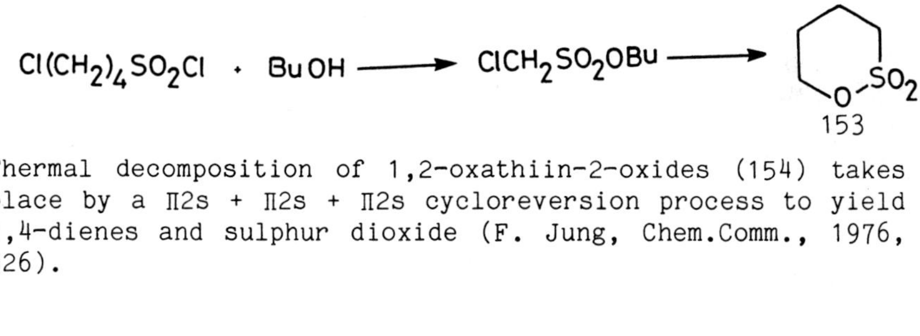

Thermal decomposition of 1,2-oxathiin-2-oxides (154) takes place by a Π2s + Π2s + Π2s cycloreversion process to yield 1,4-dienes and sulphur dioxide (F. Jung, Chem.Comm., 1976, 526).

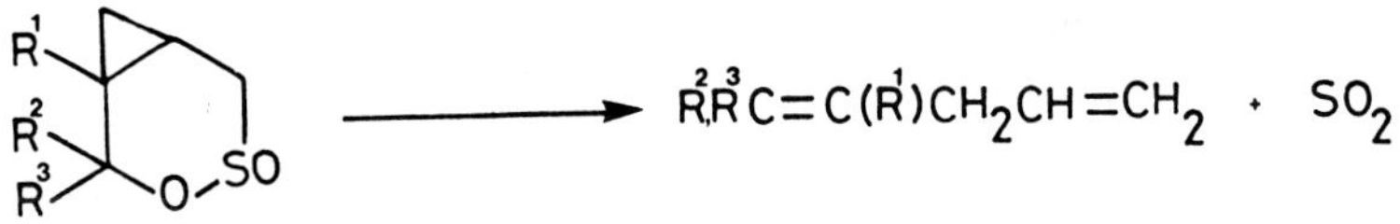

154

The reactions of 2-lithio-1,3-oxathiane with alkyl halides and carbonyl compounds has been studied and shown to give substitution and addition products as expected. Alkyl metal halides e.g. Me_3SiCl, Me_3GeCl and Me_3SnCl afford 2-trialkyl metal 1,3-oxathianes (K. Fuji et al, J.org.Chem., 1985, *50*, 657).

(ii) 1,2-Oxathiins

1,2-Oxathiin dioxides (156) are formed by the cyclo-addition of sulphene to enones (155) (A. Bargagna et al., J.heterocyclic Chem., 1980, *17*, 33), or by the addition of sulphur trioxide to dienes (T. Akiyama et al., Bull. chem.Soc.(Japan), 1978, *51*, 1251; A.V. Semenovskii et al., C.A., 1979, *91*, 123689t).

In such compounds the double bond can be epoxidised and reacted with bromine to give the 5,6-dibromo derivatives which in turn may be dehydrobrominated to dehydro 1,2-oxa-thiins (157) (Akiyama loc.cit.).

Reduction, whether chemical or electrochemical, cleaves 1,2-oxathiin dioxides to unsaturated sulphinic acids or sulphinates (S.G. Mairanoskii et al., Izv.Akad.Nauk. S.S.S.R.Ser.Khim., 1978, 2618).

(iii) 1,2-Benzoxathianes and 1,2-benzoxathiins
3-Chromanone sulphonic acid (158 ,R=OH) in the form of its sodium salt reacts with phosphorus (V) chloride to give the sulphonyl chloride (158,R=Cl), which rearranges on treatment with methylamine to the benzoxathiin dioxide (159). This product may be hydrolysed by reaction with aqueous potassium carbonate to the aldehyde (160) (W. Löwe and C.M.-Menke, Ann., 1984, 1395).

Chlorosulphene (generated from $ClCH_2SOCl_2$ and Et_3N) adds to N,N-disubstituted *E-* 2-aminomethylenecyclohexanones to afford mixtures of *cis-* and *trans-* hexahydrobenzoxathiin dioxides (161). Such mixtures may be dehydroclorinated to yield the tetrahydro analogues (162) (A. Bargagna et al., J.heterocyclic Chem., 1983, *20*, 1549).

Thiaisochroman-1,1-dioxides (164) are obtained by reacting allylbenzenes (163) with sulphuric acid (M.Ya Zarubin et al., Zh.org.Khim., 1977, *13*, 2457), whereas thienobenzoxathiin dioxides (166) are synthesised by the addition of sulphene to enones (165) (L. Mosti et al., J.heterocyclic Chem., 1982, *19*, 1227).

3,6-Dihydrobenz[b]-1,2-oxathiin-2-oxide (168) is prepared by the photochemically induced rearrangement of 1-hydroxy-1,3-dihydrobenzo[b]thiophene-2,2-dioxide (167). The product on thermolysis affords *o*-quinodimethane (169), or its equivalent, which may be trapped as adducts with α,β-unsaturated esters (J.L. Charlton and T. Durst, Tetrahedron Letters, 1984, *25*, 5287).

(b) 1,3-Oxathianes

2,2-Dialkyloxathianes (170) are synthesised by reacting 4-mercapto-2-butanol with ketones (R.A. Wilson and J.V. Pascale, (Patent), C.A., 1977, *87*, 184515x).

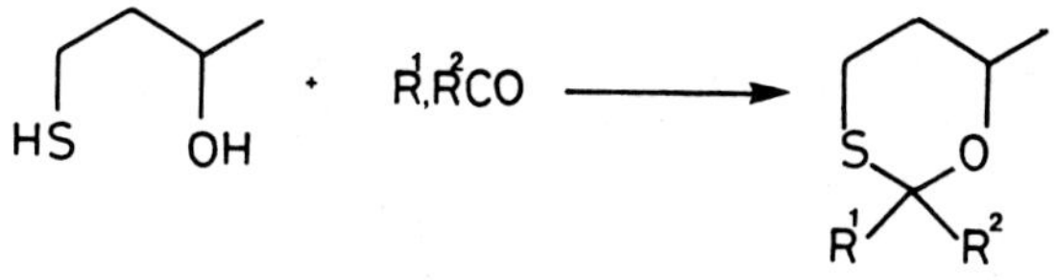

170

Monoalkylated derivatives are obtained by treatment of the salt (171) with alkyl halides (K. Fuji, M. Ueda and E. Fujita, Chem.Comm., 1977, 814; J.org.Chem., 1985, *50*, 657). Addition products form with carbonyl derivatives just as they do in similar reactions with 1,3-dithianes (see section 2b).

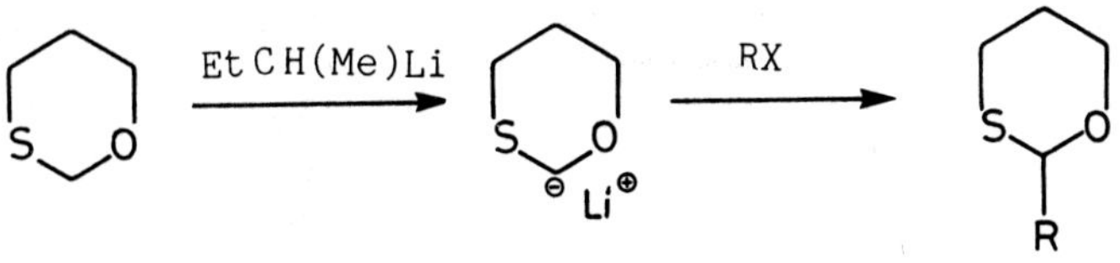

171

S-imino derivatives (172) are formed by reacting 1,3-oxa-thiane with N-chloro-amines or -amides in the presence of base. (P.K. Claus and E. Jäger, Monatsch.Chem., 1984, *115*, 1199.)

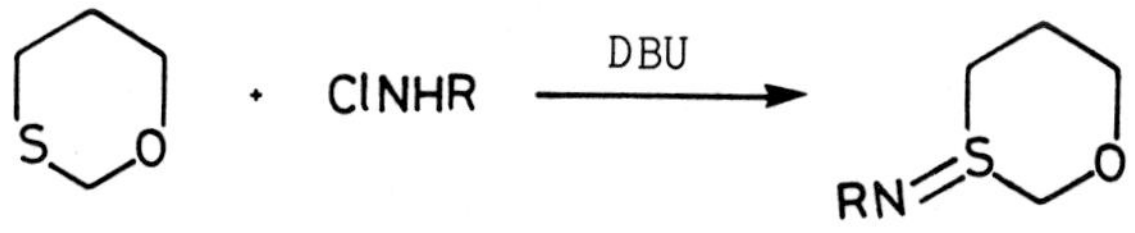

(R=Ar or ArSO$_2$)

172

2-Methyl-4-propyl-1,3-oxathiane is present in the aroma producing fraction of the yellow passion fruit, *Passiflora edulis f. flavicarpa* (G. Hensinger and A. Mosandl, Tetrahedron Letters, 1984, *25*, 507). Related structures have been made in chiral form and tested as scents and flavours (W. Pickenhagen and H. B.-Schindler, Helv., 1984, *67*, 947; A. Mosandl and G. Hensinger, Ann., 1985, 1185).

2-Acyltrimethyl-1,3-oxathianes have been studied as templates for the asymmetric synthesis of α-hydroxyaldehydes and related compounds. Thus the benzoyl derivative (173, R=Ph) when reacted with methylmagnesium iodide affords 98% of the alcohol (174) (E.L. Eliel and S.M.-Natschke, J.Amer.chem.Soc., 1984, *106*, 2937). Similarly the chiral 1,3-oxathiane (175) undergoes Grignard reactions with high stereoselectivity (J.E. Lynch and Eliel ibid., p.2943, see also K.Y. Ko, W.J. Frazee and Eliel, Tetrahedron, 1984, *40*, 1333).

173

174

175

(c) 1,4-Oxathianes, 1,4-oxathiins and related structures
(i) 1,4-Oxathianes

1,4-Oxathianes (177) are obtained by heating episulphides (176) with alcohols (V.A. Dzhafarov, Sh.K. Kyazimov and S.D. Abbasova, Dokl.Akad. Nauk.Az.S.S.R., 1977, *33*, 49). 1,4-Oxathianium salts on the other hand are formed when the ditosylate ethers (178) are treated with sodium sulphide and the product 1,4-oxathianes are then reacted with triethyl-oxonium fluoroborate (E. Kelstrup, J.chem.Soc.Perkin I, 1979, 1029).

176 + $\overset{2}{R}OH$ ⟶ 177

178

Intramolecular dehydration of the carboxylic acid sulphides (179) gives 1,4-oxathian-2-ones (180) (D.I. Davies et al. ibid., 1977, 2476). This is not the only method of preparation and J.K. Koskimies (Acta Chem.Scand., 1984, *B38*, 101) has investigated both the acid-promoted ring closure of δ-hydroxyacids, and the base-catalysted cyclisation of β-haloethyl thioglycolates (181).

179 $\xrightarrow{H^+}$ 180 $\xleftarrow{B}$ 181

1,4-Oxathian-2-one itself is completely hydrolysed back to the hydroxyacid (179,R=H) in aqueous solution at a rate comparable to that of the hydrolysis of δ-valerolactone.

(ii) 1,4-Oxathiins

1,4-Oxathiins can be synthesised by the ring-expansion of oxathialanes (182) (Patents , C.A., 1979, *90*, 103970d; 1980, *93*, 8190c; 168277g), or directly by reacting β-mercaptoethanol with chloroketo esters (183) (V.P. Arya, S.J. Shenoy and N.G. Gokhale, Ind.J.Chem., 1977, *15B*, 67).

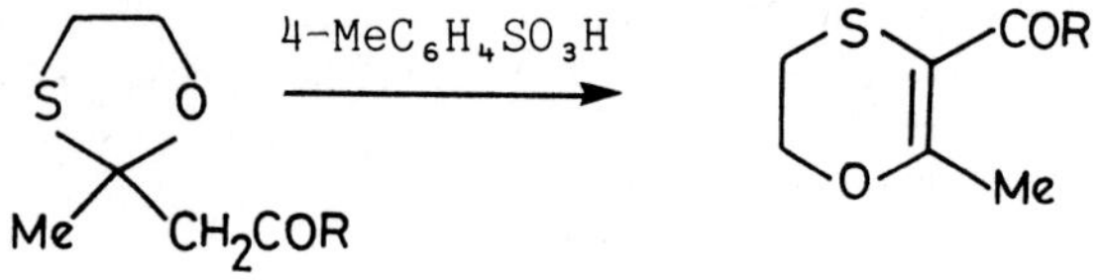

182

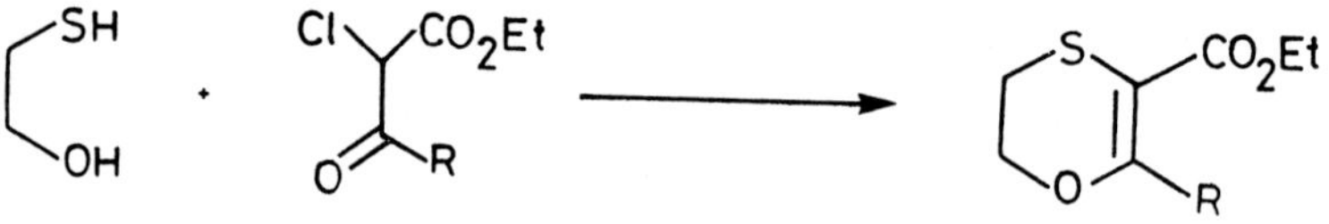

183   184

Hydrolysis of the esters (184) affords the corresponding
acids which give rise to amides when reacted with the
appropriate reagents, some of these amides are useful as
fungicides (Z. Eckstein, J. Rafa and D. Rusek, Pol.J.Chem.,
1982, *56*, 841).

When the bicyclic oxathiin (185) is heated in the presence
of methylvinylketone, the dimethylene species which is
released is trapped in the form of the adducts (186) and
(187) (B. Trost, W.C. Vladuchick and A.J. Bridges, J.Amer.
chem.Soc., 1980, *102*, 3554).

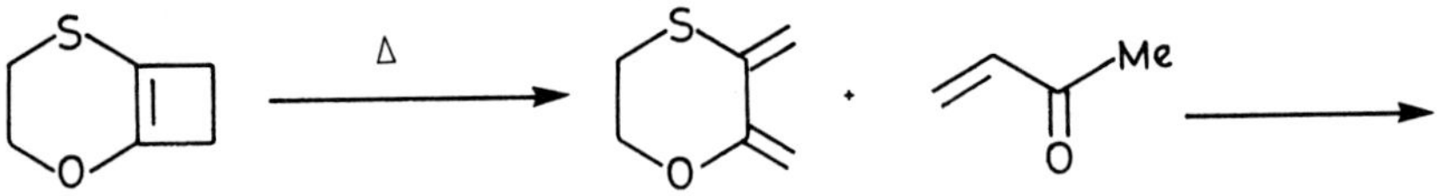

185

186   187

(iii) 1,4-Benzoxathiins

1,4-Benzoxathiins (189) are available through the cyclisation of allyl aryl ethers (188) with sulphur dichloride (M. Mühlstadt and P. Kuhl, J.prakt.Chem., 1978, *320*, 873) and 2-Alkoxycarbonyl derivatives may be synthesised by reacting 2-mercaptophenol with the dibromo esters (190) (S. Cabiddu et al., C.A., 1977, *87*, 102243v).

188   189

190

5,6,7,8-Tetrafluoro-1,4-benzoxathiin (191) is obtained in 17% yield by the reaction of potassium pentafluorothiophenate and 2-chloroethanol (Z. Domagala, R.A. Kolinski and J. Wielgat, Rocz.Chem., 1976, *50*, 993).

191

2,3-Dihydro-1,4-benzoxathiins (193) can be synthesised by treating ethylenemonothioacetals (192) of alkylcyclohexanones with copper (II) bromide (the corresponding dithiins are formed if dithioacetals are the substrates) (J.Y. Satoh et al., Chem.Comm., 1985, 1645).

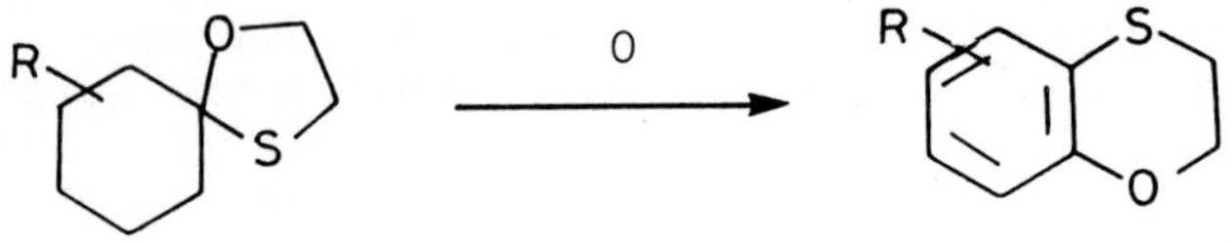

192   193

(iv) Phenoxathiins and azaphenoxathiins

1,3-Dinitrophenoxathiin (194) is the product from the reaction of 2-mercaptophenol and 2,4,6-trinitrochlorobenzene (J.C. Turley and G.E. Martin, Spectrosc.Letters, 1978, *11*, 681) – this result is a correction of earlier work reported in the literature.

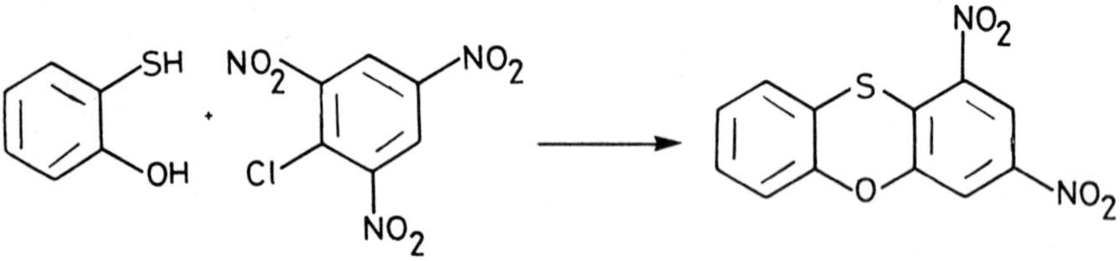

194

1-Azaphenoxathiin (195) and some analogues have been prepared from 2-mercapto-3-pyridinol disodium salt and 2-chloronitrobenzenes (Martin, Turley and Williams, J.heterocyclic Chem., 1977, *14*, 1249; 1067; 1978, *15*, 609).

1,9-Diazaphenoxathiin is synthesised by a similar route from 2-mercapto-3-pyridinol and 2-chloro-3-nitropyridine (J.S. Davies et al., Tetrahedron Letters, 1979, 5035).

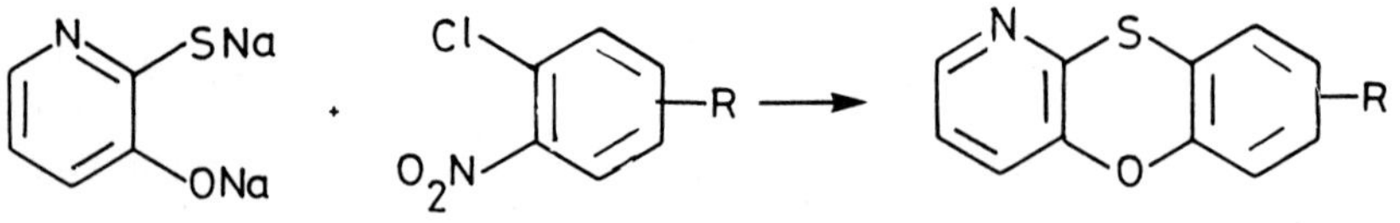

195

Chapter 41

COMPOUNDS CONTAINING A SIX-MEMBERED RING WITH TWO HETERO
ATOMS FROM GROUPS V AND VI, RESPECTIVELY OF THE PERIODIC
TABLE. OXAZINES, THIAZINES AND THEIR ANALOGUES

MALCOLM SAINSBURY

The organisation and the nomenclature used in this chapter
follow closely the systems laid down in the main work
(C.C.C. 2nd Edn., Vol. IV^H, pp. 427-535).

1.   Oxazines

(a) 1,2-Oxazines and their reduced forms

   (i) 2H-1,2-Oxazines

   These compounds are unstable and the reaction between
4-nitrosodimethylaniline (1) and tetracyclone does not give
2-(4-dimethylaminophenyl)-3,4,5,6-tetraphenyl-2H-1,2-oxazine
(2), as reported in C.C.C. 2nd edn., Vol. IV$\overline{H}$ p. 429, but
gives instead the pyrrolidinone (3) (J. Rigaudy, G. Cauquis
and J. B.-Lafont, Tetrahedron Letters, 1964, 1569).
Similarly nitrosobenzene and α-pyrone probably give the
2H-1,2-oxazine (4) as a transient intermediate, although the
isolated product is the lactam (5) (Y. Becker et al.,
J. org. Chem., 1976, *41*, 2496).

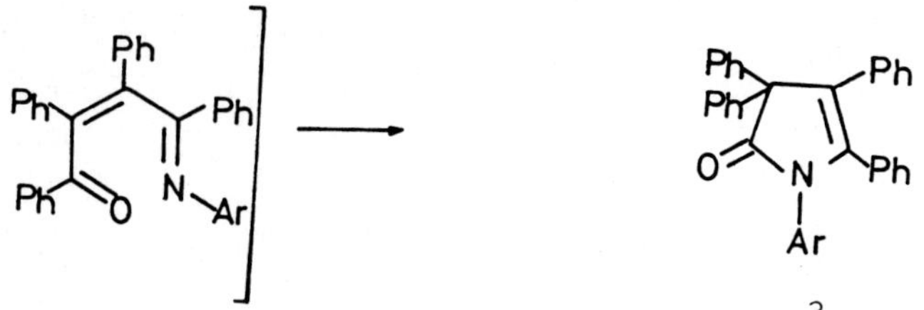

3

(Ar = 4-Me₂NC₆H₄)

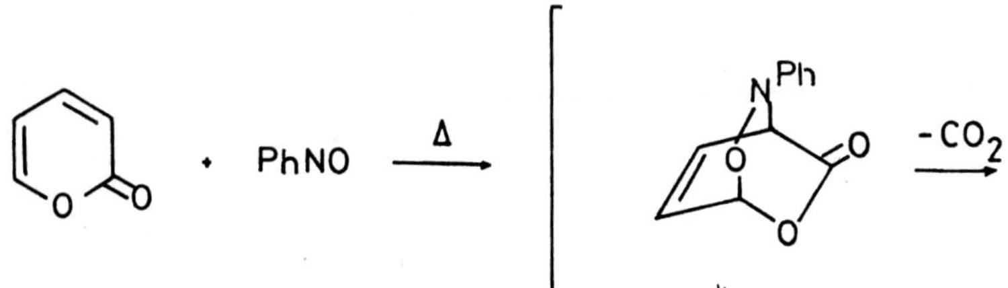

4

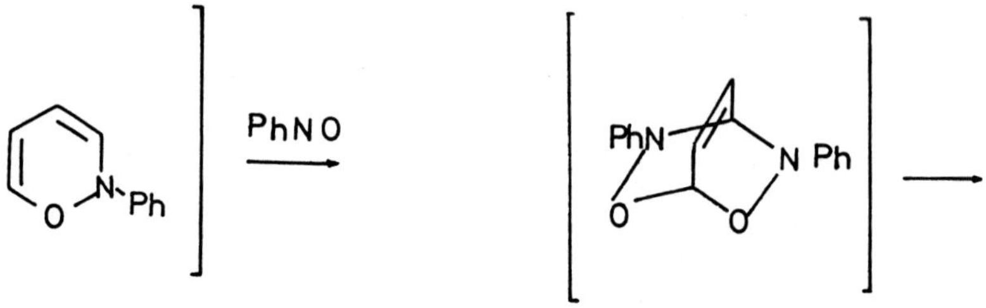

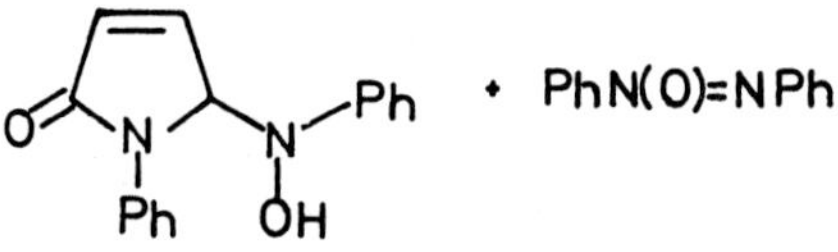

5

When *N*-hydroxycarbamic esters (6) are oxidised with tetra-
ethylammonium or sodium periodate in the presence of
conjugated dienes *N*-alkoxycarbonyl-3,6-dihydro-2H-1,2-
oxazines (7) are produced. Presumably *C*-nitrosoformate
esters are intermediates in this reaction, which then
involves cycloaddition of these species with the dienes
(G.W. Kirby et al., J.chem.Soc.Perkin I, 1985, 1437; J.E.
Baldwin, M. Otsuka and P.M. Wallace, Chem.Comm., 1985,
1549).

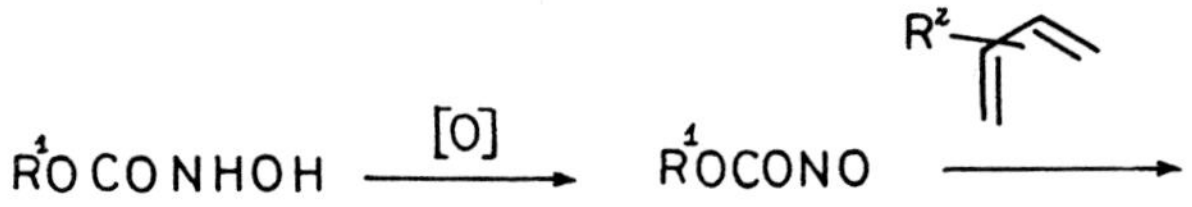

6

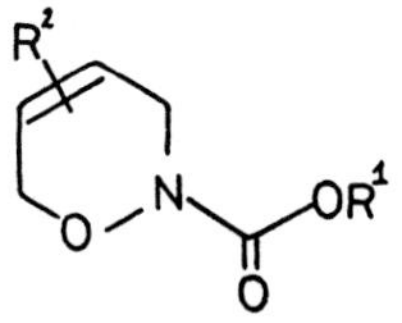

7

(ii) 6H-1,2-Oxazines

6H-1,2-Oxazines are more stable compounds than the
2H-analogues and the diesters (10) accompany the furanyl-
glyoxylate oximes (9) as products from the addition of
allylides (8) and nitrile oxides (Y. Nakada et al.,
Tetrahedron Letters, 1981, *22*, 473).

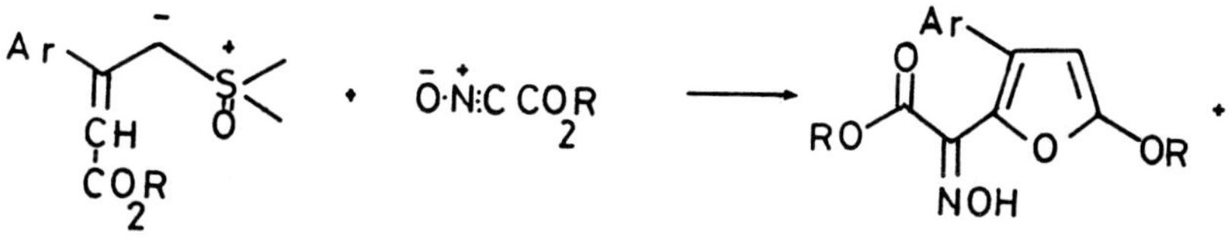

8 (R = Alkyl) 9

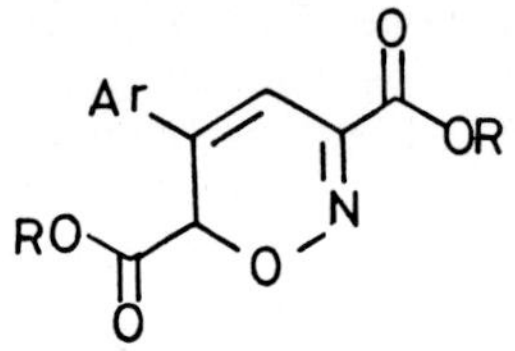

10

4-Aryl-1,2-diphenyl-2-buten-1,4-diones (11) and hydroxylamine afford 6-hydroxy-1,2-oxazines (12) (O.P. Sheyapin,
I.V. Samartseva and L.A. Pavlova, Zh.org.Chem., 1974, *10*,
1974), this being an extension of much earlier work.

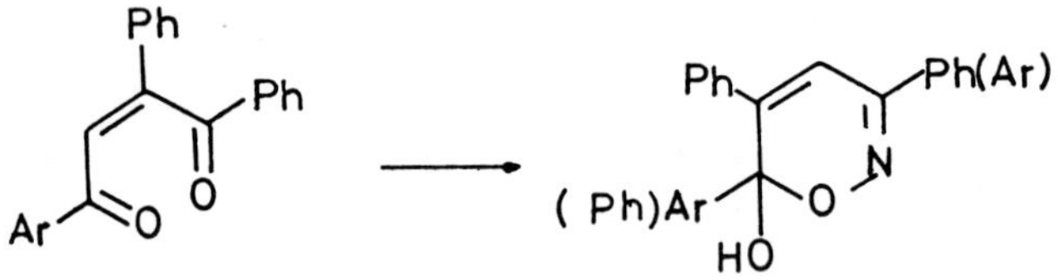

11   12

(iii) Dihydro-1,2-oxazines

3,6-Dihydro-2H-1,2-oxazines are often synthesised by the
cycloaddition of nitroso compounds and 1,3-butadienes (e.g.
see P. Häussinger and G. Kresze, Tetrahedron, 1978, 34,
689).
The parent structure (14) can be obtained in this way:
through hydrolysis of the iminium salt (13), formed by the
addition of 1-chloro-1-nitrosocyclohexane and 1,3-butadiene
(H. Labaziewicz and F.G. Riddell, J.chem.Soc.Perkin I, 1979,
2926).

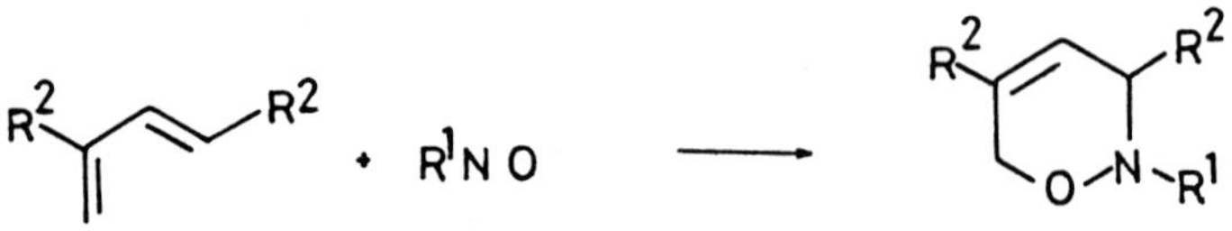

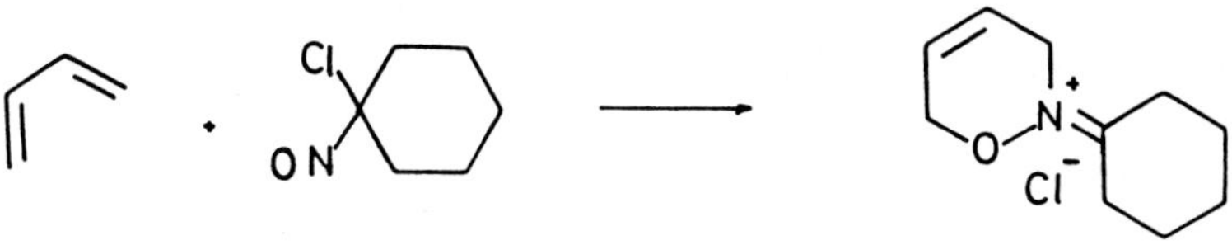

13

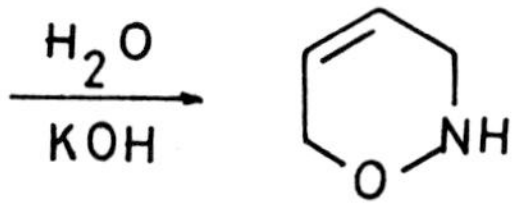

14

Nitrosoalkenes (e.g. 15) generated *in situ* from α-chloro-oximes react with alkenes to produce 5,6-dihydro-4H-1,2-oxazines (16) (T.L. Gilchrist and T.G. Roberts, Chem.Comm., 1979, 1090; R. Faragher and Gilchrist, J.chem.Soc.Perkin I, 1979, 249).

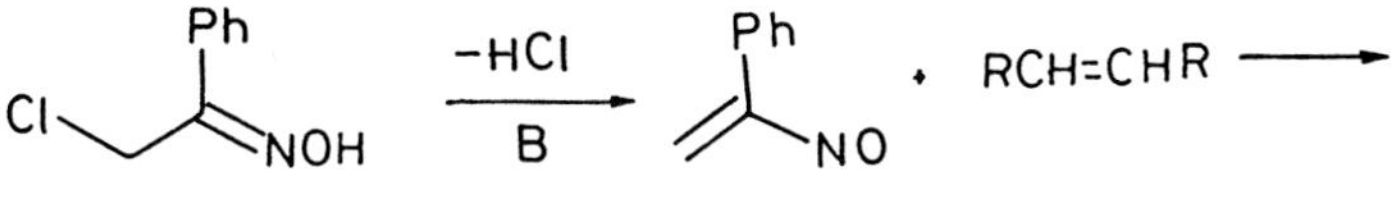

15

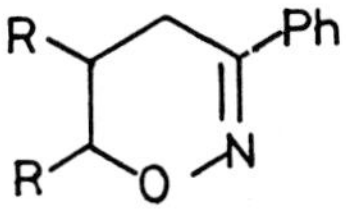

16

Similarly α-chloroacetaldehyde-*N*-cyclohexylnitrone adds to dihydrofuran and to dihydro-1,4-dioxin to afford the dihydro-1,2-oxazines (17) and (19) respectively. These products react with potassium cyanide to give the corresponding 2-cyanotetrahydro-1,2-oxazines (18) and (20) respectively (S. Shatzmiller and R. Neidlein, Tetrahedron Letters, 1976, 4151). Identical reactions occur with alkenes such as cyclohexene (E. Shalom, J.-L. Zenou and Shatzmiller, J.org.Chem., 1977, 42, 4213).

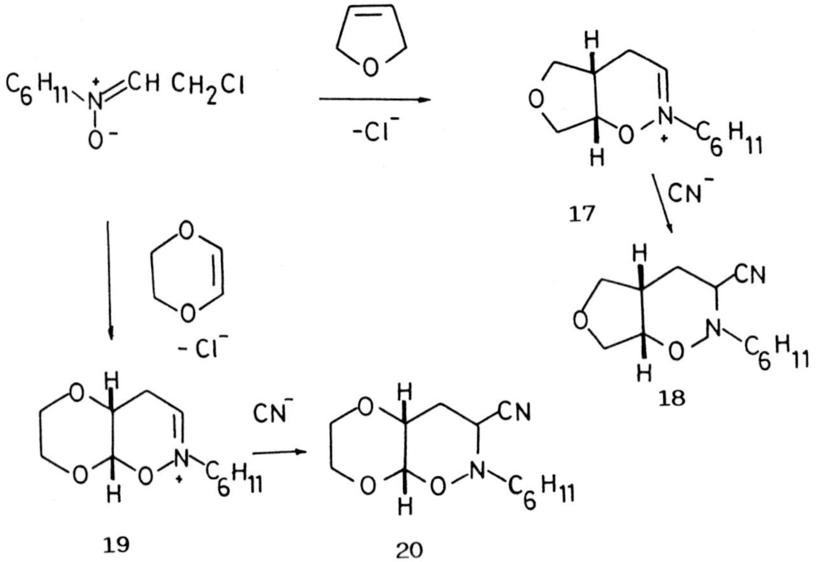

$C_6H_{11}-\overset{+}{\underset{O^-}{N}}=CH\,CH_2Cl$

$-Cl^-$

17

CN^-

18

$-Cl^-$

19   CN^- 20

The 6-nitro derivative (22) can be prepared by the cyclisation of the chloroalkyloxime (21) (P.A. Wade, J.org.Chem., 1978, 43, 2020).

Cyclopropylketones, when reacted with hydroxylamine hydrochloride, give 5,6-dihydro-4H-1,2-oxazines (23) probably through a rare example of a 6-endo tet. ring cyclisation (C.N. Rentzea, Angew Chem.intern. Edn.,1980, 92, 199). A related reaction occurs between the nitrone (24) and alkenes affording the oxazinium salts (25) (M. Riediker and W. Graf, Helv., 1979, 62, 205).

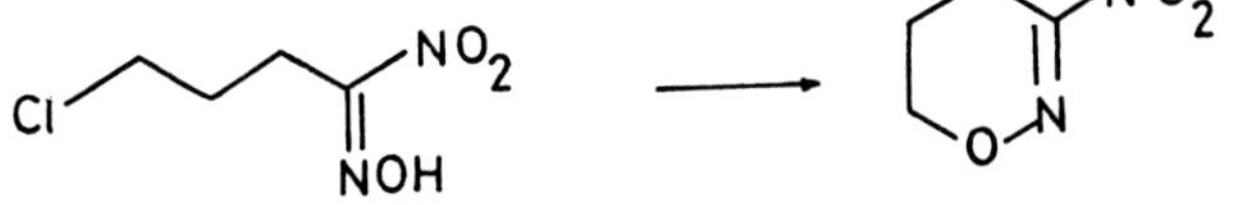

21   22

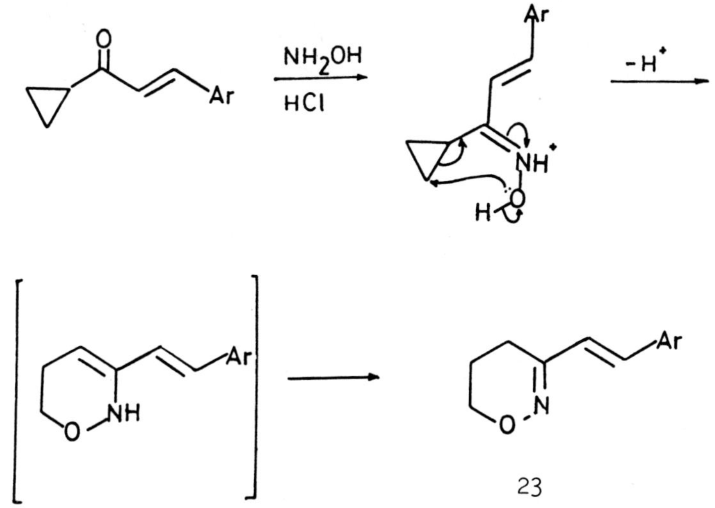

23

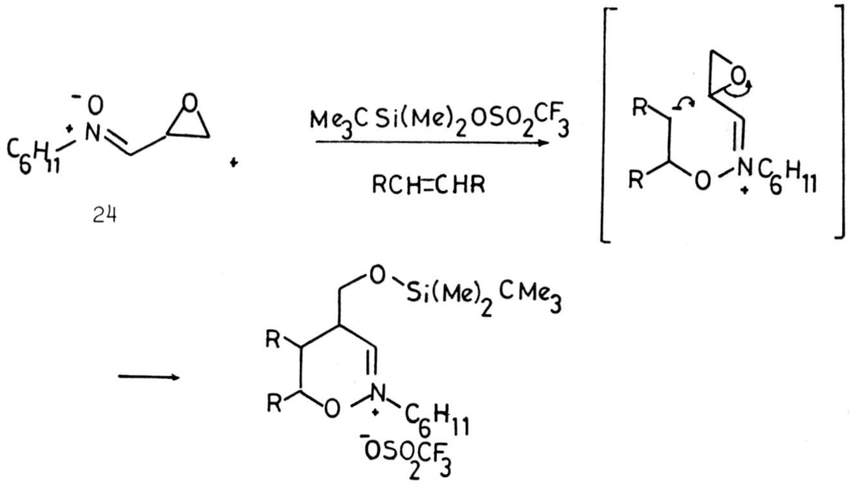

24

25

Dihydro-1,2-oxazines (27) are formed by the acid catalysed cyclisation of the carbamates (26) (B. Hardegger and S. Shatzmiller, Helv., 1976, 59, 2499).

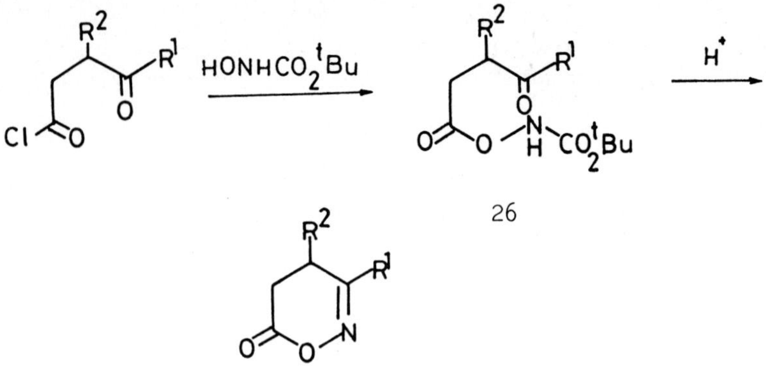

26

27

When pyrolysed the cyclopentanoxazines (28) rearrange to pyridines (29), probably through the implementation of the following steps (R. Faragher and T.L. Gilchrist, J.chem.Soc.Perkin I, 1979, 258).

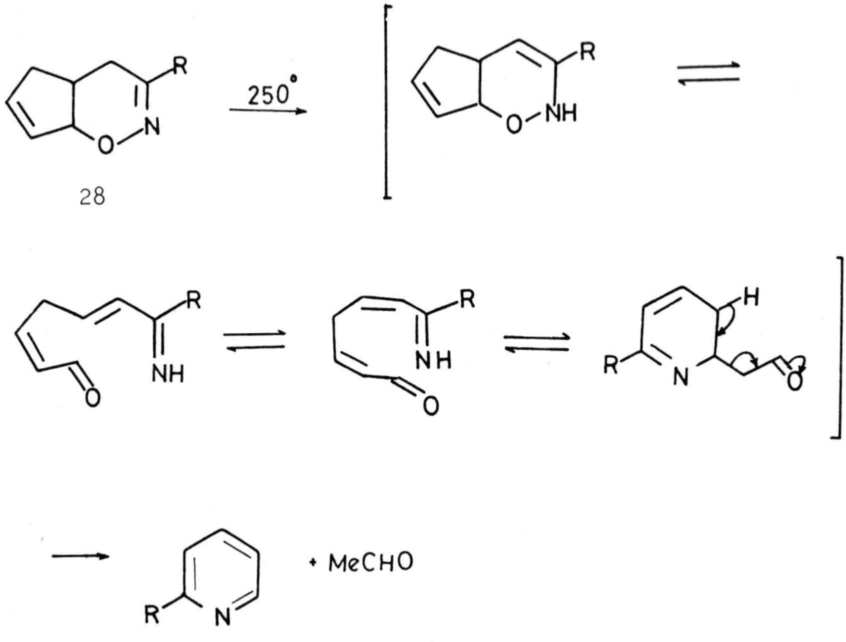

28

+ MeCHO

29

Adducts of the type (30) are cleaved with aluminium amalgam to provide amidoalcohols (31) (G.E. Keck et al., Synth.Comm., 1979, 9, 281).

(iv) Tetrahydro-1,2-oxazines

5,6-Dihydro-4H-oxazines may be N-methylated by treatment with methyl iodide and the oxazinium salts (32) which are then formed may be reduced to tetrahydro-1,2-oxazines by reaction with sodium borohydride (B. Hardegger and S. Shatzmiller, Helv., 1976, 59, 2765). Other nucleophilic reagents will react at the 3- position of the salts [e.g. KCN (see above) or BuLi], however, treatment with sodium hydride effects ring-scission.

$$CH_2=CH-C(R)=NMe$$

Tetrahydro-1,2-oxazin-3,5-diones (34) are prepared by base promoted cyclisation of the bromoketones (33). As expected on reduction with sodium borohydride the ketonic carbonyl group is transformed into a hydroxyl function but the amidic group remains unchanged (K. Tabei, E. Kawashima and T. Kato, Chem.pharm.Bull., 1979, 27, 1842; 1980, 28, 330).

Cyclohexenedicarboxylic acid anhydride (35) combines with
N-cyclohexylhydroxylamine in the presence of base to afford
the oxazin-3,6-dione (36) (P. Gyax and A. Eschenmoser,
Helv., 1977, 60, 507).

(b) 1,3-Oxazines and their reduced forms

   (i) 2H-1,3-oxazines

   New work towards 2H-1,3-oxazines is limited, however,
thermolysis of the acylazides (37) provides a route to
6-amino-1,3-oxazin-2(2H)-ones (38) (A.E. Baydar and
G.V. Boyd, Chem.Comm., 1976, 719; J.chem.Soc.Perkin I,
1981, 2871).

$$R^3_2R^3NCOC(R^2)\text{:}C(R^1)CON_3 \xrightarrow{\Delta}$$

37

38

(ii) 4H-1,3-oxazines

4H-1,3-Oxazines are synthesised by the reaction of alkynes with 1-oxa-3-azabuta-1,3-dienes, for example, phenyl-acetylene and the N—benzoylimine (39) afford 4,4-di(tri-fluoromethyl)-2,6-diphenyl-4H-1,3-oxazine (40) (K. Burger et al., Chem.Comm., 1983, 945).

39

40

Acetophenone and benzonitrile also undergo an addition reaction giving the 1,3-oxazine (41); however, in this case a Lewis acid catalyst is required (C. Kashima, S. Imad and T. Nishio, Chem.Letters, 1978, 1391).

$$PhCOMe + PhCN \xrightarrow{AlCl_3}$$

41

Nitriles react with the benzoyl derivative (42) to yield the 1,3-oxazin-4(4H)-ones (43) (E. Ziegler et al., Ann., 1977, 1751) and 1,3-oxazin-4-thiones (44) are obtained by the cyclo-condensation of acylisothiocyanates and 1,3-diketones (H. Dehne and P. Krey, J.prakt.Chem., 1982, 324, 915).

$$42 \quad + RCN \longrightarrow 43$$

$$R^1CONCS + R^2COCH_2COR^3 \longrightarrow 44$$

1,3-Oxazin-4-ones undergo a number of ring-contraction reactions giving other heterocycles. For example, with hydrazines 6-methyl-2-phenyl-1,3-oxazin-4(4H)-one (45) affords triazoles (46) (Y. Yamamoto, Y. Azuma and K. Miyakawa, Chem.pharm.Bull., 1978, *26*, 1825), hydroxyl-amine yields 1,2,4-oxadiazoles (47) (Yamamoto and Azuma, Heterocycles, 1978, *9*, 185). On the other hand, the anions of lactams give α-substituted lactams (48) (Yamamoto and Y. Morita, Chem.pharm.Bull., 1984, *32*, 2555).

$$46 \qquad 45 \qquad 47$$

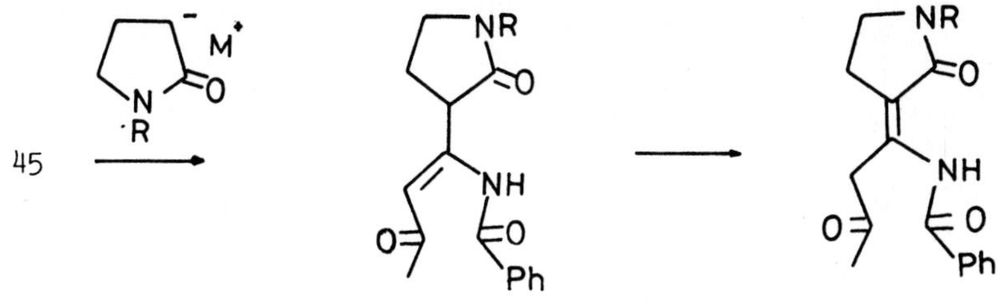

45

48

With enolate anions however ring contraction does not occur, now initial ring-opening is followed by recyclisation to afford pyridones (idem. ibid., 1985, *33*, 975).

When isoxazol-5-ones are treated with phenylcyanate ring-expansion to 1,3-oxazin-6-ones (49) occurs (F. Risitano et al., J.chem.Soc.Perkin I, 1979, 1522).

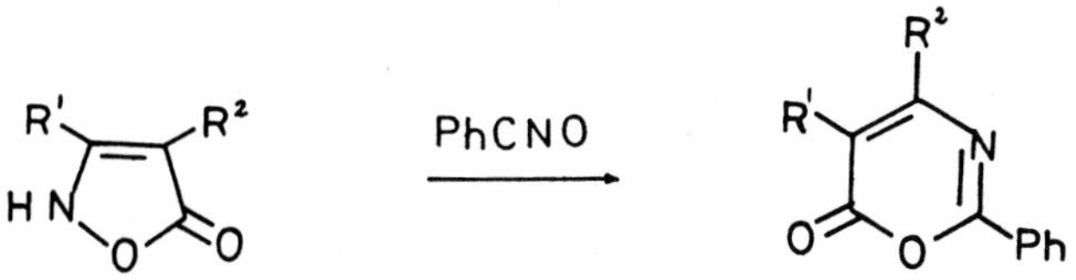

(49)

Similar compounds are also synthesised by the action of zinc and acyl chlorides on α-bromoacetates and nitriles (H. Lapkin, V.I. Semenov and M.I. Belanovich, Zhur.org. Khim., 1977, *13*, 1328); and thienooxazines (51) are obtained by the cyclisation of thiophene carboxylates (50) (I.A. Kharizomenova et al., Khim.Geterot.Soedin., 1980, 45).

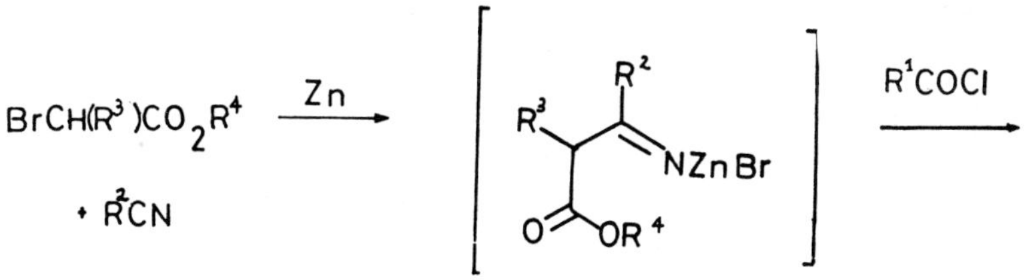

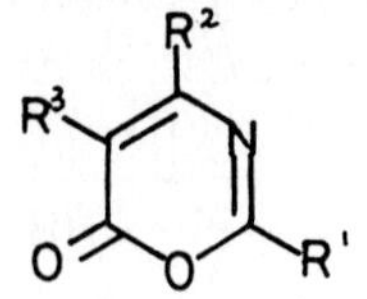

50

51

On flash-vacuum pyrolysis 1,3-oxazin-6-ones fragment to carbon dioxide, alkynes and nitriles (P.W. Manley et al., Chem.Comm., 1978, 902), but on irradiation with ultra-violet light valence tautomers (52) are produced, which then lose carbon dioxide to give azetes (53) (P. DeMayo, A.C. Weedon and R.W. Zabel, ibid., 1980, 881; G. Maier and V. Schaefer, Ann., 1980, 798).

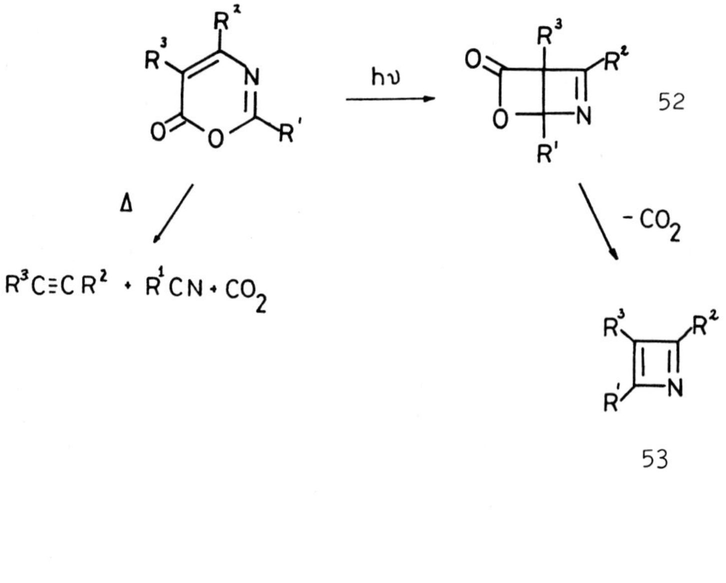

53

Norbornenes (55) at 150° undergo retro-Diels-Alder reactions to give 6H-1,3-oxazin-6-ones (56) (G. Stajer et al., Synthesis, 1984, 345). The starting materials are prepared from the amides (54) by treatment with thionyl chloride and triethylamine.

(iii) 1,3-Oxazinium salts and mesoionic structures

N-Acyl-β-enamino ketones (57), prepared from N-(substituted) dithiocarbonimidates and the potassium enolates of methyl ketones, on treatment with perchloric acid or methanesulphonic acid yield 1,3-oxazinium salts (58) (K.T. Potts, A.J. Ruffini and G. Titus, J.org.Chem., 1983, *48*, 623).

Mesoionic 1,3-oxazines (60) are formed by reacting secondary amides with malonyl chlorides (59) (W. Friedrichsen et al., Ann., 1978, 1655). These compounds combine with

3,4,5,6-tetrachloro-1,2-benzoquinone to produce tricyclic adducts (61) (idem., Tetrahedron Letters,1979, 237).

R^2NHCOR^1 + $R^3CH(COCl)_2$ ⟶

59

60

61

(iv) Dihydro-1,3-oxazines

Most interest in this area centres around 4H-5,6-dihydro-1,3-oxazines. These compounds can be prepared in a variety of ways, for example, by:(a) the addition of nitriles to oxetanes (62) (G. Schneider, L. Hacklerand, P. Sohar, Tetrahedron Letters, 1981, *22*, 341; T.M. Pavel, Zhur.org.Khim., 1982, *18*, 178); (b) the acid mediated cyclisation of 1,3-amido alcohols (63) (A.P. Gusaev et al., C.A., 1985, *101*, 130645d); (c) the interaction of β-hydroxyalkenes with nitriles in the presence of concentrated sulphuric acid (A.A. Gevorkyan and G.G. Tokmadzhyan, (Patent), C.A., 1979, *90*, 204113g); and (d) the cycloaddition, in two steps, of allenyl nitriles and 3-aminopropanol (S.R. Landor et al., Heterocycles, 1981, *16*, 1889).

+ $\overset{1}{R}CN$ ⟶

62

Some other routes involve (a) inverse demand Diels Alder reactions between triethyl azomethinetricarboxylate (64) and alkenes bearing electron-donating substituents (H.K. Hall Jr., and D.L. Miniutti, Tetrahedron Letters, 1984, 25, 943); (b) the cyclocondensation of carboxylic acids and aminoalcohols in the presence of triphenyl-phosphine, pyridine and carbon tetrachloride (H. Vorbrueggen and K. Krolikiewicz, ibid., 1981, 22, 4471); and(c) the reaction of aminoalcohols with thioamide salts (T. Harada, Y. Tamaru and Z. Yoshida, Chem.Letters, 1979, 1353).

$(EtO_2C)_2C=N\,CO_2Et$ + $R^1CH=CR^2R^3$ ⟶

64

2-Cyano-3-phenylpropionic, or 2-bromo-2,3-dicyanopropionic esters undergo self-condensation in the presence of phosphites to yield 4,5-dihydro-2H-1,3-oxazines (65) (F. Texier, E. Marchand and A. Foucaud, Compt.rend., 1975, 281C, 51). Additionally, the 3,4-dihydro-2H-1,3-oxazin-2-one (66) is formed by the reaction of chlorosulphonyl isocyanate with chalcone (D.N. Dhar, G. Mehta and S.C. Suri, Ind.J.Chem., 1976, 14B, 477).

1,3-Dioxinones (67) react with imines to afford 2H-3,4-dihydro-1,3-oxazin-4-ones (68) and with isocyanates to give oxazindiones (69) (M. Sato, H. Ogasawara and T. Kato, Chem.pharm.Bull.(Japan), 1984, *32*, 2602).

68   67   69

Imines also enter into addition reactions with unsaturated lactones (70) to produce dihydro-1,3-oxazin-4-ones (71) (Yu S. Andreichikov et al., Khim.Geterot.Soedin., 1978, 271).

70   71

5,6-dihydro-1,3-oxazin-5(4H)-ones (73) are uncommon, but may be obtained by the treatment of diazoketones (72) with mineral acids (V.G. Kartsev and A.M. Sipyagin, ibid., 1980, 565; 1324).

72   73

Ethyl acetoacetate and *N,N*-dimethylurea react in acetic anhydride and acetic acid solution to give 6-methyl-1,3-oxazine-2,4(3H)-dione. The same compound is also formed by heating ethyl *N*-acetoacetylcarbamate, or by reacting

N'-acetoacetyl-N,N-dimethylurea with sulphuric acid (S. Ahmed, R. Lofthouse and G. Shaw, J.chem.Soc.Perkin I, 1976, 1969).

The 5-methyl analogue (75) is obtained by the acid mediated cyclisation of the vinyl ether (74) (J. Farkas, Coll.Czech.chem.Comm., 1979, 44, 269), and the N-silylated derivative (76) of the parent structure is formed by reacting maleic anhydride with trimethylsilylazide at 55-60° (J.H. Macmillan, Org.prep.Proced.Int., 1977, 9, 87). On treatment with ethanol the N-silyl bond is cleaved.

This last compound is an inhibitor of *Escherichia coli* and other bacteria. It may be halogenated directly at C-5 by chlorine, bromine or iodine monochloride (Farkas, O. Fliegerova and J. Skoda, Coll.Czech.Chem.Comm., 1976, *41*, 2059), and its derivatives undergo deuterium exchange at this site (S.S. Washburne and H. Lee, Tetrahedron Letters, 1978, 1693).

Alkyl halides in the presence of triethylamine alkylate the sulphur atom of the 2-thione (77), but the products (78) are then easily cleaved by potassium carbonate to alkyl thiocyanates and acetylketene (Y. Yamamoto and Y. Morita, Chem.pharm.Bull., 1984, *32*, 2957).

77   78

Bicyclic structures (79) obtained by the addition of chloroformylketenes to 2-pyridone, enter into addition reactions with phenyl isocyanate and also with acetylene dicarboxylates (T. Kappe et al., Ber., 1979, *112*, 1585).

79

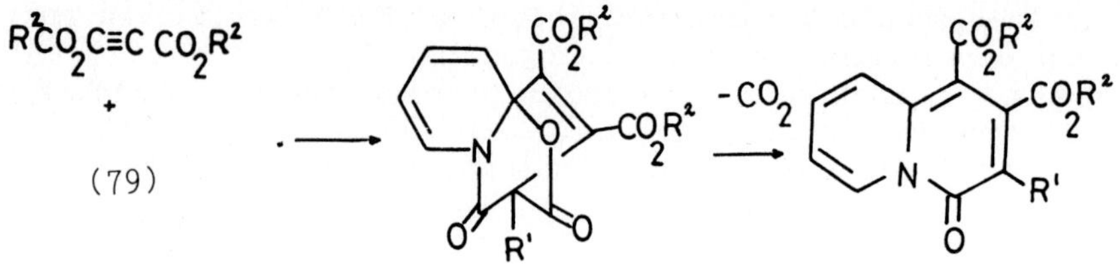

$$R^2O_2C \equiv C \; CO_2R^2$$

$$+$$

(79)

(v) Tetrahydro-1,3-oxazines

Tetrahydro-1,3-oxazin-2-ones (81) are formed by the addition of oxetanes (80) and isocyanates in the presence of the complex of dibutyltindiiodide and triphenylphosphine oxide (A. Baba et al., Tetrahedron Letters, 1985, *26*, 5167) and the 5-nitrotetrahydro-1,3-oxazine (83)is obtained from 2-(hydroxymethyl)-2-nitro-1,3-propandiol (82) by treatment with 1,3,5-trimethylhexahydro-s-triazine. When reacted with metal alkoxides it yields the nitronate (84) through a retro-aldol process (M.P. Crozet and P. Vanelle, ibid., p.323).

80   + R^2NCO $\longrightarrow$ 81

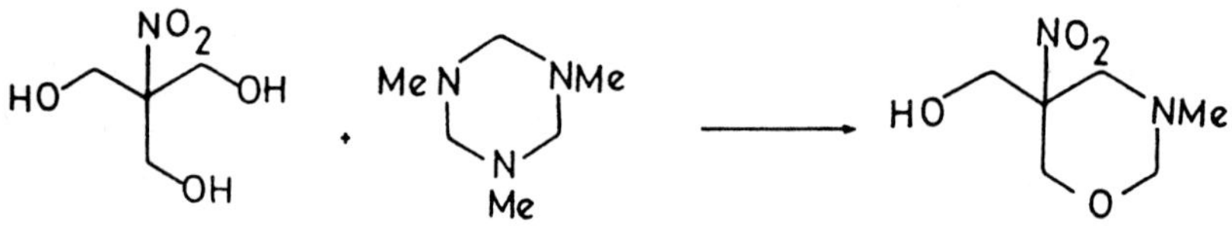

82   83

MOMe

$\longrightarrow$

(M= Na or Li)

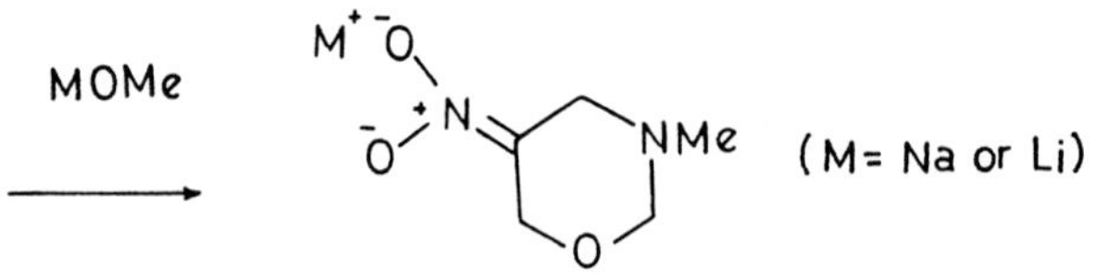

84

When tetrahydro-1,3-oxazine is allowed to stand for several weeks it trimerizes to the triazine (85); however, the process may be reversed by heating this product at approximately 125°C.

85

Sequential treatment of tetrahydro-1,3-oxazines (86) with diketene and then with 4-toluenesulphonylazide/triethyl-amine gives rise to the diazoketone (87), which on irradiation with ultraviolet light yields the β-lactams (88) (R.J. Ponsford and R. Southgate, Chem.Comm., 1979, 846).

86   87   88

Tetrahydro-1,3-oxazin-2-ones are usually prepared by the cyclisation of γ-amino alcohols with ethyl chloroformate phosgene or similar reagents (see, for example, T. Urbánski and A.S.-Szalowska, Bull.Acad.Pol.Sci.Ser. Chem., 1976, 24, 447; E. Abignente, M.P.-Biniecka, Acta.Pol.Pharm., 1979, 36, 511). The preparation of the corresponding 2-thiones employs thiophosgene or carbon disulphide as reagents (A.S. Moskovkin et al., C.A., 1983, 98, 160655u).

Methylation of tetrahydro-1,3-oxazin-2-thiones with methyl iodide affords salts, which with sodium hydroxide give the corresponding 2-ones (A.B. Khasirdzhev and B.V. Unkovskii, C.A., 1977, 86, 72541a).

Tetrahydro-1,3-oxazin-4-ones are formed by the reaction of ketones (T. Kametani et al., Heterocycles, 1977, *7*, 919; 1978, *9*, 819), or aldehydes (F. Fulop et al., J.chem.Soc.Perkin I, 1984, 2043) with β-hydroxyamides.

N-Alkoxy-3-hydroxypropionamides and aryl isocyanates cyclise in the presence of triethylamine to produce 2,4-dioxotetrahydro-1,3-oxazines (D. Geffken, Ber., 1979, *112*, 600).

If thiophosgene is used instead of the isocyanates the corresponding 2-thiones are obtained (Geffken, Chem.Ztg., 1979, *103*, 79).

(c)   1,4-Oxazines

(i) Dihydro-1,4-oxazines

2H-3,4-Dihydro-1,4-oxazines are synthesised by treating ketals (89) with phosphorus (V) oxide in pyridine (P.W. Freeman, S.N. Quessy and L.R. Williams, Synth.Comm., 1979, *9*, 631).

Cyclisation of the hydrobromides of amino ketoesters (90) is effected by molecular sieves or by hot triethylamine. In either event the products are

the   dihydro-1,4-oxazin-2-ones   (91)   (G.   Schulz   and
W. Steglich, Ber., 1977, *110*, 3615;   V. Caplar et al.,
J.org.Chem., 1978, *43*, 1355).

89

90   91

N-Phenacylglycinates on treatment with 4-toluenesulphonic
acid ring-close to give related structures (92) in which
there is a Δ^5-bond (M. Benet and R. Longeray, Compt.rend.,
1979, *288C*, 229).

92

(ii) Tetrahydro-1,4-oxazines (morpholines)

Sequential treatment of ethanolamines (93) and oxiranes
with 70% sulphuric acid and hydrogenolysis over palladium on
carbon provides a route to 2,3-disubstituted tetrahydro-
1,4-oxazines (G. Bettoni et al., Tetrahedron, 1980, *36*, 409;
see also F.Loftus, Synth.Comm., 1980, *10*, 59).

2-Aryl derivatives are formed if the haloethers (94) are
treated with arylmagnesium bromides, followed by cyclisation
of the products (95) with amines (N. Busch et al.,
Eur.J.med.Chem.Chim.Ther., 1976, *11*, 201).

Tetrahydro-1,4-oxazinones arise through (a) the reaction of oxiranes with α-amino acids, and (b) the combination of ethanolamines and α-bromoacetates (A. LeRouzic, D. Raphalen and M. Kerfanto, Compt.rend., 1979, *288C*, 221).

Ethanolamines, ketones and chloroform in the presence of sodium hydroxide also afford tetrahydro-1,4-oxazin-2-ones (J.T. Lai, Synthesis, 1984, 122), whereas tetrahydro-1,4-oxazin-3-ones (96) can be obtained by the oxidation of morpholines with ruthenium (VIII) oxide (R. Perrone, G. Bettoni and V.Tortorella, ibid., 1976, 598; Perrone, F. Berardi and Tortorella, Gazz., 1983, *113*, 521).

N-Formylmorpholine reacts with alkyl lithiums or alkyl-magnesium bromides to give aldehydes. (G. Olah, L. Ohannesian and M. Arvanaghi, J.org.Chem., 1984, *49*, 3856). 3-Methoxy-4-formylmorpholine also acts as a formylating agent for arenes in the presence of aluminium (III) chloride (L. Eberson, M. Malmberg and K. Nyberg, Acta Chem.Scand., 1984, *B38*, 345).

Glyoxal reacts with 2-(*N*-methylamino)ethanol at room-temperature to afford the dicyclic and tricyclic structures (97) and (98) respectively. The relative stereochemistries of these products were deduced from X-ray crystallographic analyses (A. Le Rouzic, M. Maunaye and P. L'Haridon, J.chem.Res. Sp., 1985, 35).

97

98

2. Benzoxazines

(a) 1,2-Benzoxazines

(i) 1H-2,1-Benzoxazines
N-Acyl-3,4-dihydro-1H-2,1-benzoxazines are obtained through the cyclisation of *N*-halo-*N*-alkoxyamides (99) in the

presence of Lewis acids.   *N*-acyl *N*-alkoxynitrenium ions are likely intermediates in these reactions (S. Glover   et al., J.chem.Soc.Perkin I, 1984, 2255).

(ii) 1H-2,3-Benzoxazines

Representatives of this structural class are known in the form of 3,4-dihydro derivatives usually bearing a carbonyl group at position 1.   Such compounds (e.g.100) are prepared from 2-chloromethylbenzoyl chlorides by reaction with hydroxyl-amines (G. Zinner and V. Ruthe, Chem.Ztg., 1977, *101*, 451; A. Trani and E. Bellasio, J.heterocyclic Chem., 1977, *14*, 113).

Dehydro compounds (102) are synthesised by reacting carbonylazides (101) with hydroxylamine in pyridine solution, or from *N*-hydroxyphthalimides and arenes by treatment with aluminium (III) chloride (A.F.M. Fahmy, N Aly and M.Oraby, Egypt.J.Chem., 1976, *19*, 223).   The 1-phenyl derivative (102,Ar=Ph), first made by M.P. David and J.F.W. McOmie (Tetrahedron Letters 1973, 1361), on flash vacuum pyrolysis rearranges to the 3,1 -benzoxazine (103) by a C-N transposition mechanism (K.L. Davies, R.C. Storr and P.J. Whittle, Chem.Comm., 1978, 9).

Reduction of these oxazin-1-ones with sodium borohydride or
with lithiumaluminium hydride yields the alcohols (104), but
aryllithiums effect both arylation and ring-cleavage
(E.B. D'yakoskaya et al., Zh.org.Khim., 1978, *14*, 1701;
L.A. Pavlova, I.V. Samartseva and N.V. L'vova, ibid., 1979,
15, 2405).

(b) 1,3-Benzoxazines and reduced forms

(i) 2H-1,3-Benzoxazines
4-Chloro-2H-1,3-benzoxazines (105) are useful starting materials for a number of other heterocycles, for example, when reacted with pyridine-*N*-oxides they afford *N*-2-pyridyl derivatives which when hydrolysed give 2-aminopyridines (K. Wachi and A. Terada, Chem.pharm.Bull., 1980, *28*, 465). Reactions with 2-aminophenols yield 2-hydroxyphenzyl-benzoxazoles (106) (R. Tachikawa, Wachi and Terada, Heterocycles, 1981, *16*, 1165) and similarly 2-aminoaceto-phenone produces the quinazoline (107) (idem., Chem. pharm.Bull., 1982, *30*, 559).

(ii) 1,3-Benzoxazin-2-ones and 4-ones

1,3-Benzoxazin-2-ones are available through the reactions of 2-hydroxyphenyl ketones with chlorosulphonyl isocyanate (A. Kamal and P.B. Sattur, Synth.Comm., 1982, *12*, 157), and by treatment of the imines (108) with *N,N*-carbonyldiimidazole (P. Stoss, Ber., 1978, *111*, 314). If thiophosgene is used in the last reaction 1,3-benzoxazin-2-thiones are formed.

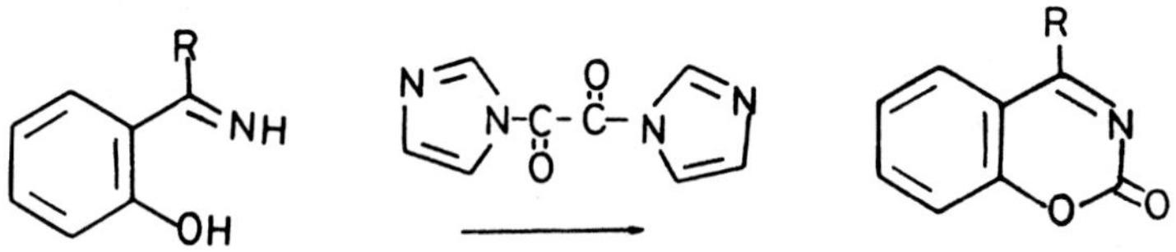

108

2-*N,N*-dialkylamino-1,3-benzoxazin-4-ones can be synthesised by treating 2-hydroxybenzonitriles with phosgeniminium chlorides (e.g.109) (B. Kokel, G. Menichi and M. H.-Habart, Tetrahedron Letters, 1984, *25*, 3837).

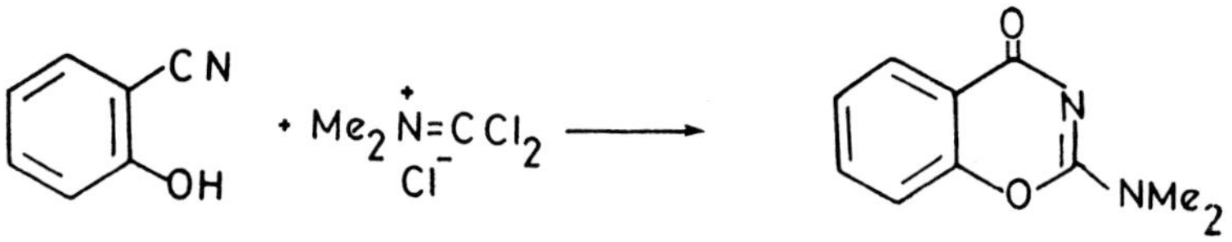

109

(iii) Dihydro-1,3-benzoxazines

Dihalides (110) undergo cyclocondensation with primary amines to afford 3,4-dihydro-2H-1,3-benzoxazines (111) (C.L. Colin and B. Loubinoux, ibid., 1982, *23*, 4245).

110   111

(iv) Dihydrobenz-1,3-oxazinones and thiones

3,4-Dihydro-1,3-benzoxazin-2(2H)-ones (113) are easily made from 2-hydroxybenzylamines (112) by reaction with potassium isocyanate (J. Arct, E. Jakubska and G. Olszewska, Synth.Comm., 1977, *8*, 143).   The corresponding thiones arise if the reagent used is potassium isothiocyanate (Arct and Olszewska (Patent), C.A., 1981. *94*, 4024z).
2-Hydroxybenzamides and ethyl orthoformate give 2-ethoxy-1,3-benzoxazin-4-ones (114) (S. Treppendahl and P. Jakobsen, Acta Chem.Scand., 1981, *B35*, 465), whereas the base catalysed ring-expansion of benzisoxazolinones (115) yields 2,2-disubstituted derivatives (116) (H. Uno and M. Kurokawa, Chem.pharm.Bull., 1978, *26*, 549).

112   113

114

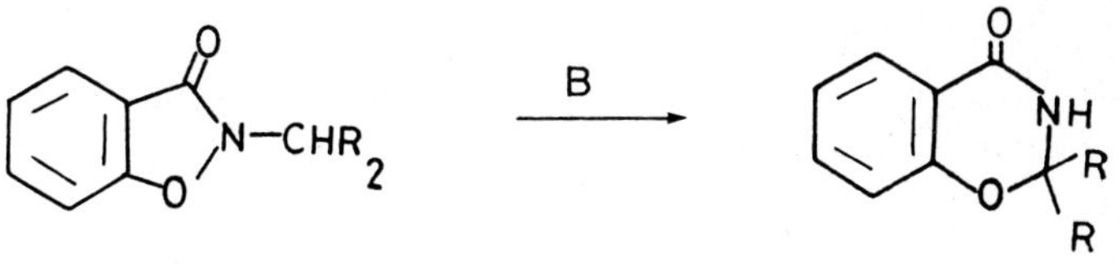

115 → 116

(v) 1,3-Benzoxazindiones and related compounds

1,3-Benzoxazin-2,4-diones are well established compounds and are normally synthesised from 2-hydroxybenzamides and reagents such as ethyl chloroformate (G. Wagner, D. Briel and S. Leistner, Pharmazie, 1980, *35*, 49). 3-Amino derivatives are prepared by treatment of the benzoyl-hydrazones (117) with this reagent (J.S. Davidson, Chem.Ind., 1982, 660).

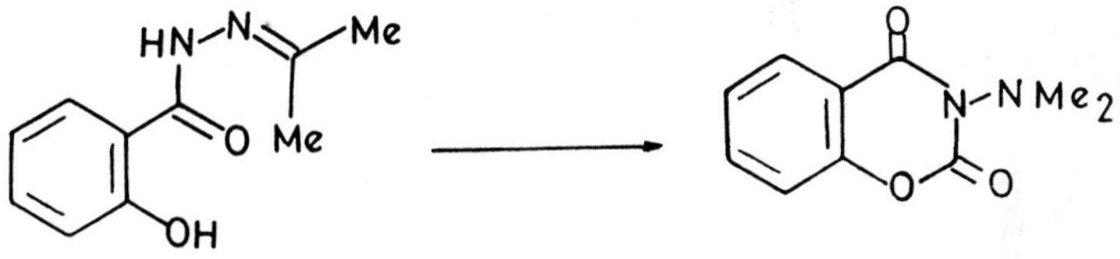

117

2-Thion-4-ones (118) are generated from salicylic acids by the action of triphenylphosphine diisothiocyanate, at low temperature(Y. Tamura et al., ibid., 1978, 806), and 4-imino-2-ones arise through the reaction of isocyanates and 2-hydroxybenzonitrile in the presence of triethylamine (J.P.-Fischer and E.P. Papadopoulus, J.heterocyclic Chem., 1983, *20*, 1159).

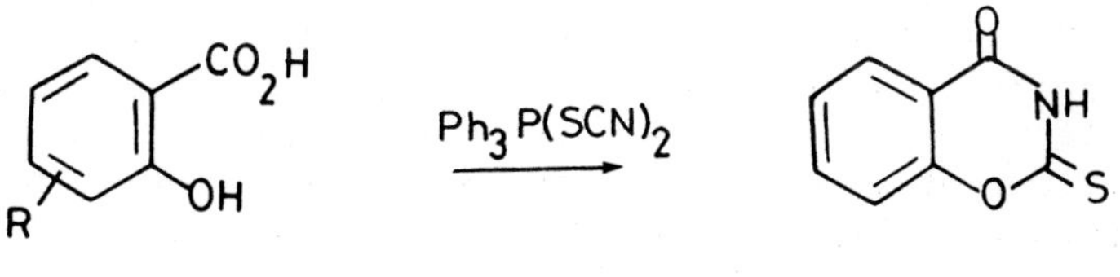

118

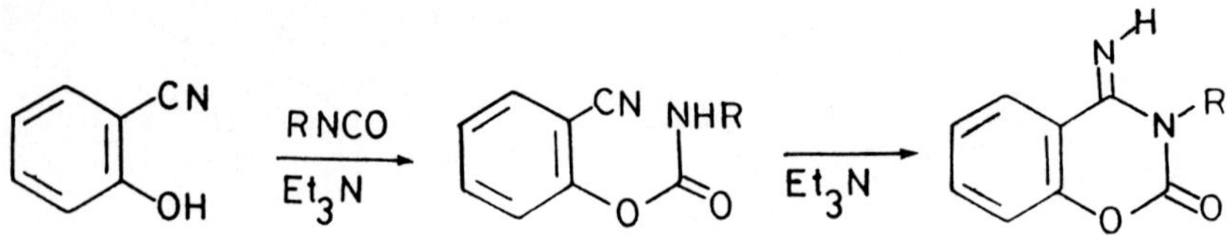

(c) 3,1-Benzoxazines and reduced forms

(i) 4H-3,1-Benzoxazines

A synthesis of 4H-3,1-benzoxazines (120) requires the reaction of phenyl isocyanates with primary or secondary amines. The initial products are ureas (119) which then cyclise on heating (W. Gauss and H.J. Kabbe, Synthesis, 1978, 377).

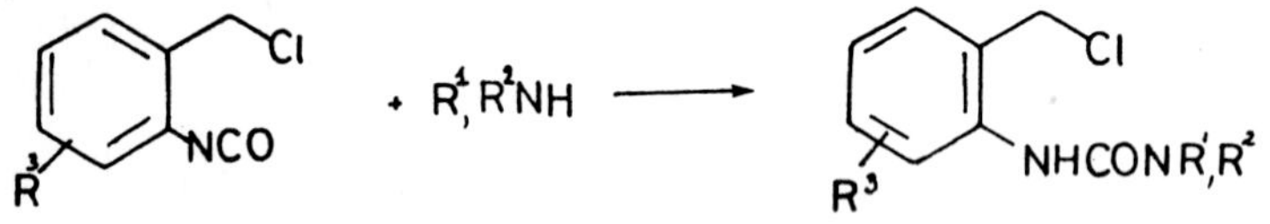

119

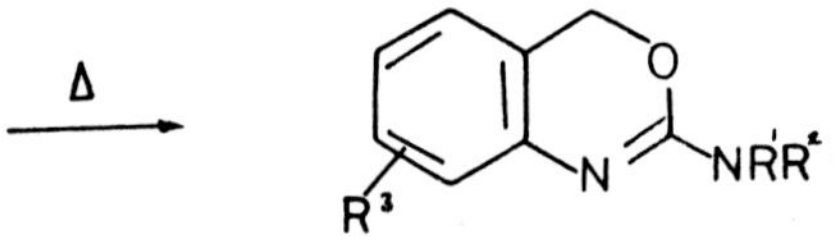

120

(ii) 3,1-Benzoxazin-4-ones

These are the most numerous representatives of this heterocyclic system and they are almost invariably derived from anthranilic acid or its derivatives. Thus N-benzoyl-anthranilic esters are cyclised by treatment with cold sulphuric acid (E.P. Papadopoulos and C.D. Torres, Heterocycles, 1982, 19, 1039); N-acylanthranilic acids are ring-closed with triphenyl phosphite and pyridine (G. Rabilloud and B. Sillion, J.heterocyclic Chem., 1980,

17, 1065), with acetic anhydride or phosphorus oxychloride and pyridine (Z. Ecsery et al., (Patent) C.A., 1979, *91*, 39500s), or with thionyl chloride (H. Nohira et al., Heterocycles, 1977, *7*, 301). Treatment of indol-3(3H)ones (121) with hydrogen peroxide affords hydroperoxides (122) which then rearrange to 3,1-benzoxazin-3-ones (J.M. Adam and T. Winkler, Helv., 1984, *67*, 2186).

121   122

With amines 3,1-benzoxazin-4-ones ring-open to yield 2-acylaminobenzamides (I. Bitter, L. Szocs and L. Toke, Acta chim.Acad.Sci.Hung., 1981, *107*, 57), and with zinc and acetic acid are cleaved to afford *N*-alkylanthranilic acids (B.M. Bolotin et al., Khim.Geterot.Soedin., 1977, 1619).

(iii) 2,4-Dihydro-1H-3,1-benzoxazines and related compounds

A simple synthesis of 1,4-dihydro-2H-3,1-benzoxazines (123) is furnished by the cyclo-oxidation of 2-(*N*-methyl-amino)benzyl alcohols by reaction with manganese (IV) oxide (F. Kienzle, Tetrahedron Letters, 1983, *24*, 2213).

123

An approach to fused systems (125) involves the photochemical cyclisation of the 2-(*N*-pyrrolylamino)benzyl
alcohol (124) (F.Z. Basha and R.W. Franck, J.org.Chem.,
1978, *43*, 3415).
A Baeyer-Villiger reaction upon isatins (126) using
persulphuric acid in sulphuric acid yields 1,4-oxazin-
2,3-diones (127), but with hydrogen peroxide in acetic acid
the isatoic anhydrides (128) are obtained (G. Reissenweber
and D. Mangold, Angew Chem., 1980, *92*, 196).

(d) 1,4-Benzoxazines and reduced forms

2-Aminophenols and α-keto aldehydes cyclodehydrate to yield
2H-1,4-benzoxazines (129) (E. Belgodere et al., J.hetero-
cyclic Chem., 1980, *17*, 1625), however, α-ketoesters afford
1,4-benzoxazin-2-ones (M.T. LeBris, J.heterocyclic Chem.,
1984, *21*, 551). Another route to 3-aryl derivatives uses
phenacylbromides (D.R. Shridhar et al., Synthesis, 1981,
913). Similarly 3-alkyl-2H-1,4-benzoxazines (131) are
formed by the cyclisation of acetamidoketones (130) in the
presence of 6N hydrochloric acid (F. Chioccara and
E. Novellino, J.het.Chem., 1985, *22*, 1021). The 3-methyl
compound is very unstable, whereas its *t*-butyl analogue
undergoes slow aerial oxidation to the hemiacetal (132).

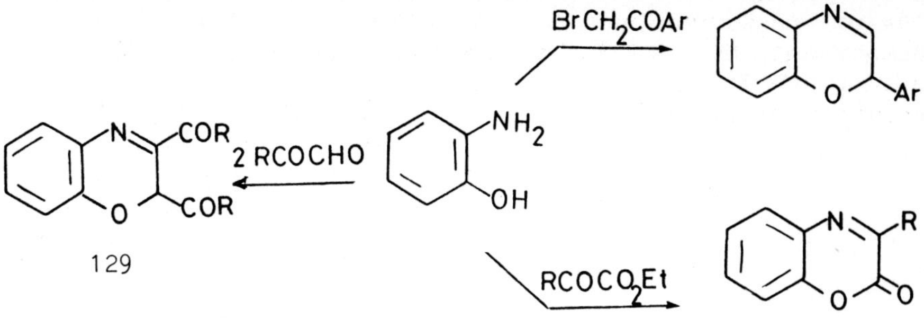

129

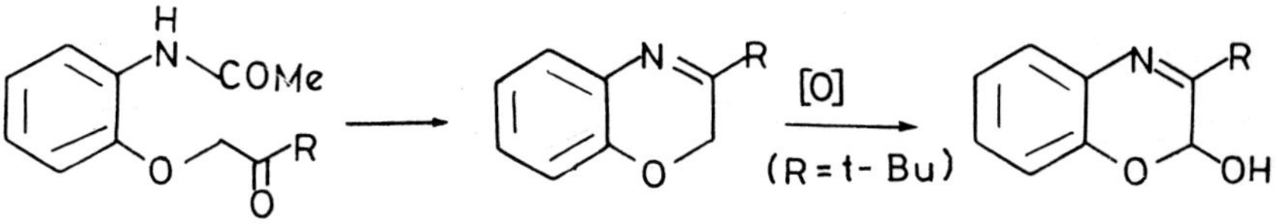

130   131   132

N-Oxides (134) are available through the partial reduction of 2-nitro ethers (133) with zinc dust and ammonium chloride (P. Battistoni, P. Bruni and G. Fava, Tetrahedron, 1979, *35*, 1771;   see also H. Bartsch, W. Kropp and M. Pailer, Monatsch., 1979, *110*, 257).   These products undergo cyclo-addition with acrylonitrile to yield tricycles (135).

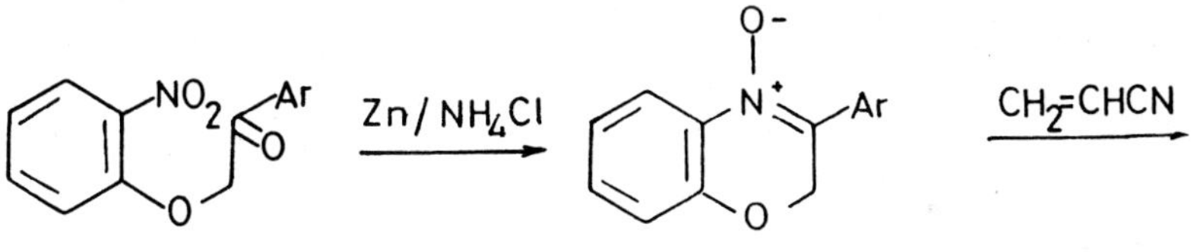

133   134

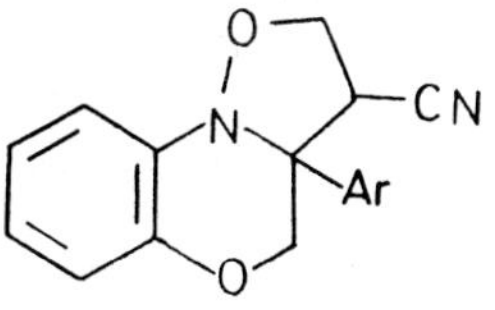

135

On irradiation with ultraviolet light 1,4-benzoxazin-2-ones
undergo photodimerisation (T. Nishio and Y. Omote,
Chem.Comm., 1984, 1293), but when electron poor alkenes are
present, such as vinylcarboxylates or cyanoolefins, these add
across the imine bond (idem., J.org.Chem., 1985, *50*, 1370).

Those derivatives formally having a substituted phenacyl
group at position-3 preferentially exist in the enamine form
(136) with internal hydrogen bonding (Y. Iwanami and
I. Inagaki, J.heterocyclic Chem., 1976, *13*, 681).

136

Pyrolysis of the benzoxazine derivative (137) affords the
tricyclic compounds (138) and (139). The last structure is
assumed to be an intermediate *en route* to a further product
— the dimer (140) (N. Kawahara and T. Shimamori, Hetero
cycles, 1985, *23*, 2401).

2H-1,4-Benzoxazin-3(4H)- ones (141,X=O) are readily available
from 2-aminophenols and chloroacetyl chloride
(D.R. Shridhar, M. Jogibhuknan and V.S.H. Krishnan,
Org.prep.Proced.Int., 1982, *14*, 195; and the corresponding

thiones (141,X=S) are obtained from them by treatment with phosphorus (V) sulphide (idem. ibid., 1984, *16*, 91).

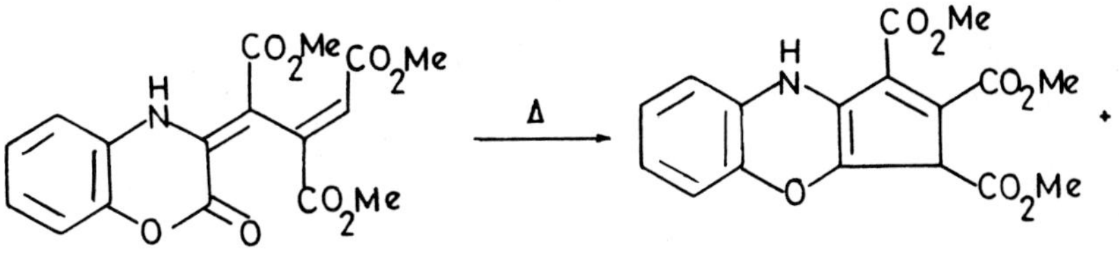

137

138

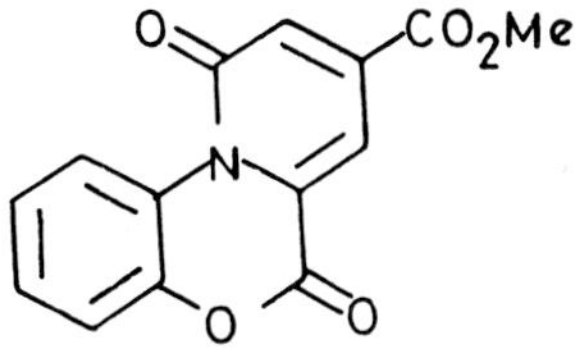

139

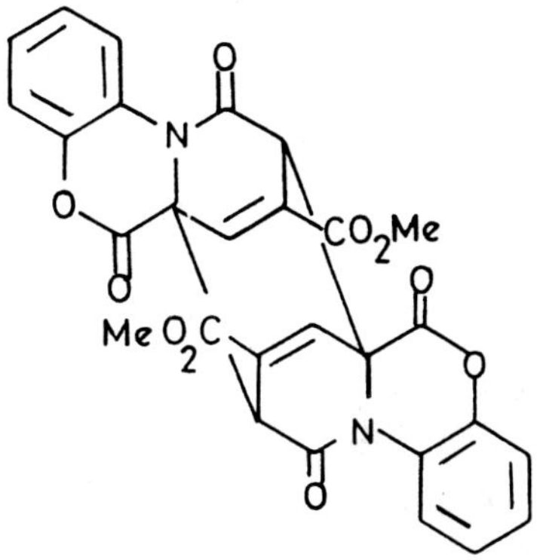

140

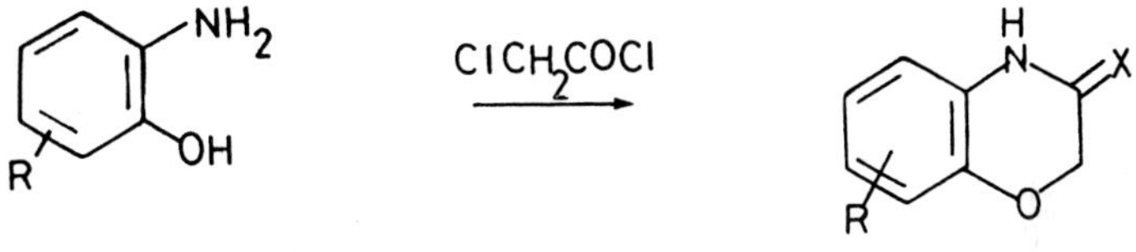

141

2-Hydroxy-*N*-propargylanilines in the presence of mercury (II) oxide give 2-methylene-3,4-dihydro-2H-1,4-benzoxazines (142) (M. Yamamoto et al., J.chem.Soc.Perkin I, 1979, 3161), and the parent heterocycles (143) are obtained by reacting 2-hydroxycarboxyanilides with dibromoethane in aceto-nitrile/dichloromethane solution containing powdered sodium hydroxide

142

143

3.   Phenoxazines

3,7-Disulphonylated phenoxazines arise through the reaction of phenoxazine with sodium sulphinates in acetic acid containing iron (III) chloride (L.S. Kapishenko, A.V.Prosyanik and S.I. Burmistrov, Khim.Geterot.Soedin., 1977, 616). Presumably cation radical intermediates are involved although when these are deliberately formed and reacted with various nucleophiles 3-substituted phenoxazines result (H.J. Shine and S. Wu, J.org.Chem., 1979, 44, 3310). Oxidation of 3-nitrophenzoxine occurs on irradiation in air with sunlight yielding phenoxazin-3-(3H)-one (H. Musso, Ber., 1978, 111, 3012).

A new approach to such products involves the cyclisation of nitro ethers with zinc dust and ammonium chloride. The reaction mixture is aerated (C.W. Bird and M. Latif, Tetrahedron, 1980, 36, 529).

Cinnabarinic acid (144) is obtained in high yield (90%) by the oxidative ring closure of 2-amino-3-hydroxybenzoic acid with manganese (IV) oxide (W. Prinz and N Savage, C.A., 1978, *88*, 22792d).

144

2-Substituted phenoxazin-3(3H)-ones react with amines in boiling alcoholic solution to yield 3— amino compounds (145). However, 1,2-phenylene diamine gives the pentacycles (146) (G.B. Afanas'eve, I.Ya Postovskii and T.S. Viktorova, Khim.Geterot.Soedin., 1978, 1200).

145   146

In the case of 2-dialkylamino-3H-phenoxazin-3-ones (147, R=alkyl) photolysis with a high pressure lamp effects cleavage of one of the alkyl groups (Y. Ueno, Ann., 1983, 161).

Reduction of diacylphenoxazin-3(3H)-ones (148) with sodium dithionate in the presence of diethyl sulphate affords the corresponding ethoxy derivatives (149), together with the pentacyclic structures (150) (A. Bolognese, C. Piscitelli and G. Scherillo, C.A., 1983, *101*, 211065s), similar polycyclic molecules (151), as well as angular structures (152), are formed by the action of dilute methanolic sulphuric acid on the phenoxazinones (idem., J.org.Chem., 1983, *48*, 3649).

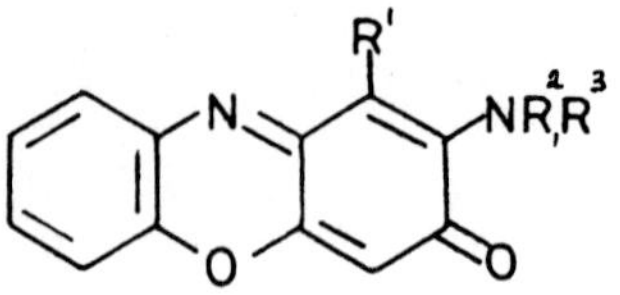

147

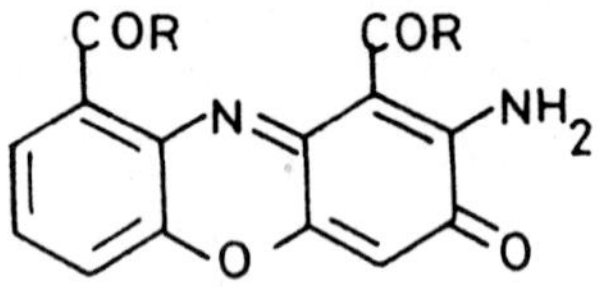

148

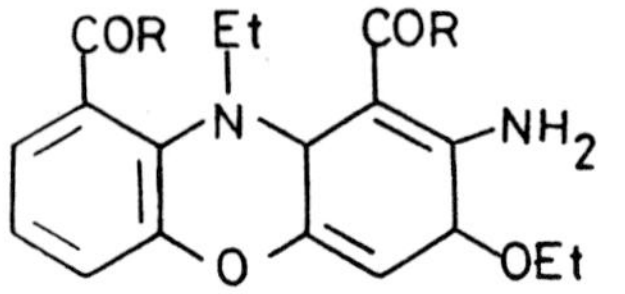

149

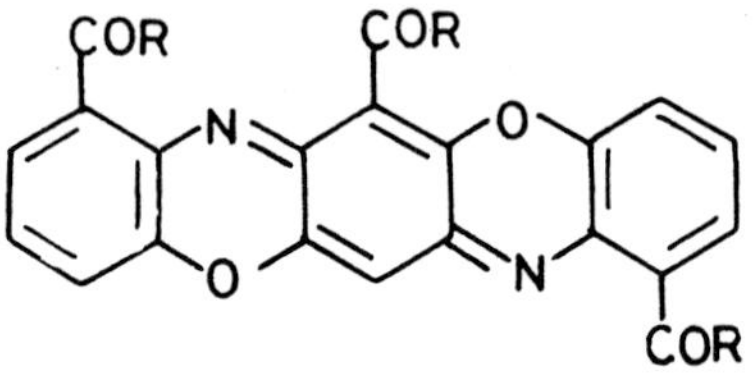

150

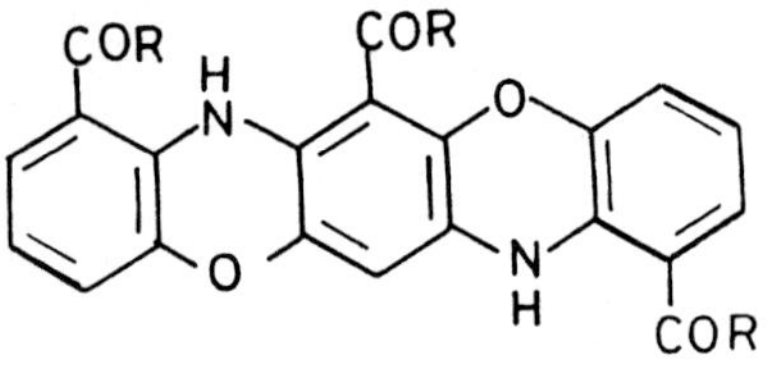

151

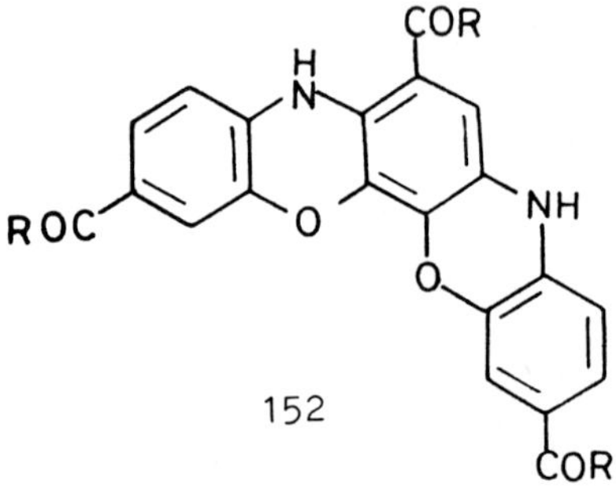

152

Other fused systems (154) are available through the interaction of the naphthoquinone (153) and 2-aminophenols (N.L. Agarawal and W. Schäfer, ibid., 1980, *45*, 2155). It appears that these products form as a result of attack by two molecules of the aminophenol, followed by loss of one of them, rather than by a reaction involving a Smiles rearrangement, as might have been predicted.

An unexpected product from the nitrosation of 2-amino-phenoxazin 3-one is oxadiazolo[5,4-a]phenoxazin-4-one (155) (A. Küchen, P. Bigler and U.P. Schunegger, Chimia, 1984, *38*, 387).

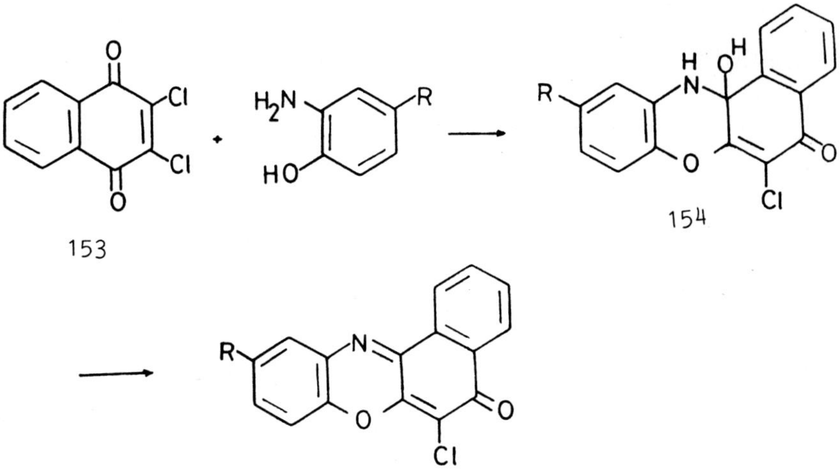

153

154

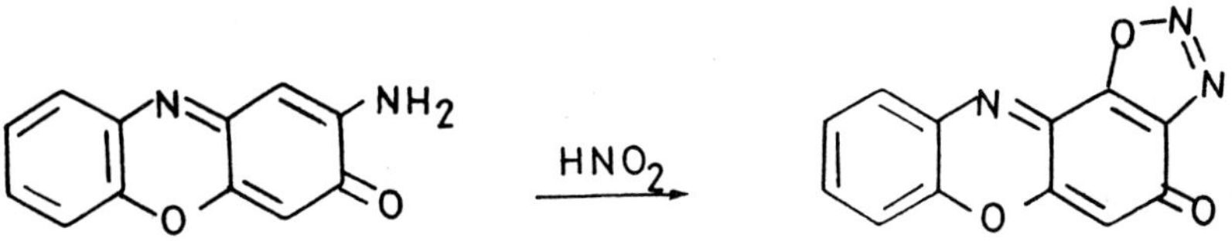

155

4. Thiazines

(a) *1,2-Thiazines and reduced forms*

When the main work was published in 1978 1,2-thiazine derivatives were uncommon. Since then a number of representative structures have been synthesised, for example, the perfluorothiazine (156) is available through the cycloaddition of perfluoro-1,3-butadiene and thiazyl fluoride (W. Bludssuns and R. Mews, Chem.Comm., 1979, 35). Similarly several *S*-oxides (158) have been obtained through the base catalysed cyclisations of *N*-(2,2-diacylvinyl) dimethylsulphoximines (157) (Y. Tamura et al., J.org.Chem., 1977, *42*, 602; M. Ieda et al., Heterocycles, 1983, *20*, 2185; see also W.D. Rudorf, Synthesis, 1983, 926).

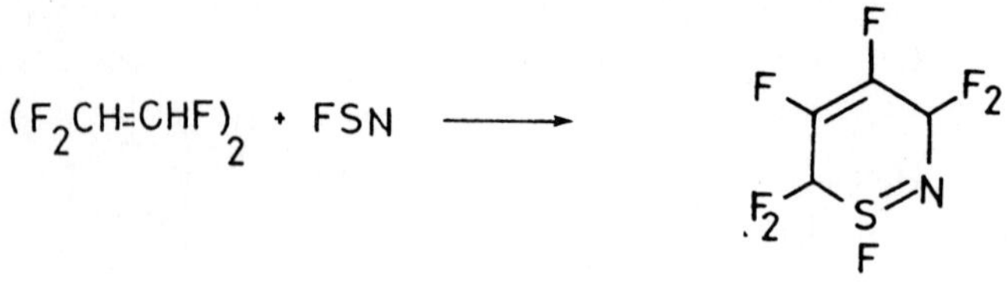

$(F_2CH=CHF)_2$ + FSN ⟶

156

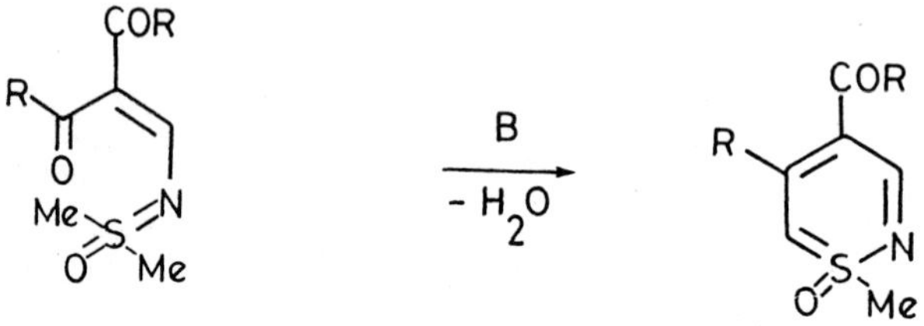

157

B
$- H_2O$

158

N-Sulphinyliminium salts (159) combine with dienes to produce 2 <u>H</u>-3,6-dihydro-1,2-thiazines (160) (G. Kresze and M. Rossert, Angew Chem., 1978, *90*, 61), and related structures (162) are obtained by the addition of sulphin - imines (161) to dienes (E.S. Levchenko and E.I. Slyusarenko, C.A., 1977, *87*, 23185z).

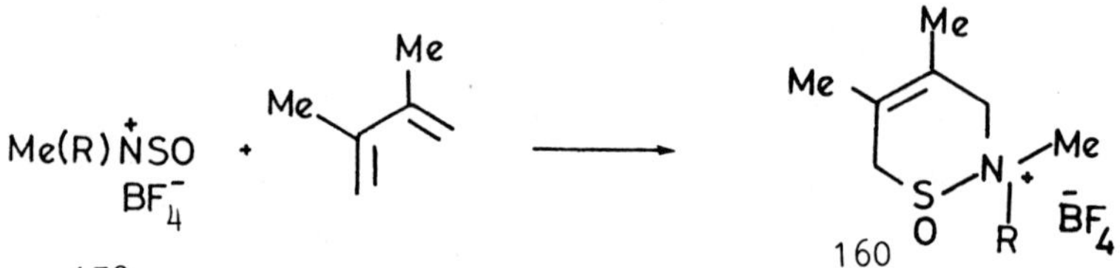

$Me(R)\overset{+}{N}SO$ +
BF_4^-

159

160

ArNSO +

161

162

1,2-Thiazin-5(6H)ones (166) are formed by the addition of sulphonylamides (163) to O-silylated dihydroxydienes (164), followed by deprotection with hydrochloric acid. A certain amount of the diketosulphonamides (165) are isolated in the reaction, but on heating these cyclise to the thiazinones (J.A. Kloek and K.L. Leschinsky, J.org.Chem., 1979, 44, 305).

Reduced forms are synthesised by reacting arylamines with δ-chloroalkylsulphonyl chlorides (S.H. Doss, A.B. Sakla and M. Hamed, Org.Prep.Proced. Int., 1978, 10, 48), and the oxidised structures (167) may be formylated under Vilsmeier conditions (DMF and POCl$_3$) at position C-6 (E. Fanghaenel, A.M. Richter and H. Muhammmed, (Patent) C.A., 1985, 102, 62253t). Such compounds react with chlorine in a rather indiscriminate manner, ultimately yielding the pentachloro-derivatives (168) (Fanghaenel et al., J.prakt.Chem., 1984, 326, 545).

(b) 1,3-Thiazines and related compounds

(i) 1,3-Thiazines

4H-1,3-Thiazines (169) and/or their tautomers (170) are available through the cyclisation of amido ketones with phosphorus (V) sulphide (T. Hirayama et al., (Patent), C.A., 1979, *90*, 103982j). Representative structures of this type are also prepared by the addition of the acetylene (172) to selenothiazoles (171) (K. Burger and R. Ohlinger, Synthesis, 1978, 44).

169   170

171   172

6H-1,3-Thiazines can be synthesised in a number of ways. Thus 3-chloroalkenylisothiocyanates (173) can be cyclised by treatment with amines or with alcohols (K. Schulze et al., J.prakt.Chem., 1984, *326*, 101); the thionoimines (174) undergo cycloaddition with α,β-unsaturated ketones (J.C. Meslin and H. Quiniou, Bull Soc.chim.Fr., 1979 *7-8*, 347; T.C. Gokou, J.P. Pradere and Quiniou, J.org.Chem., 1985, *50*, 1545); and enolate amination of the phosphonate (175) with O-mesitylenesulphonylhydroxylamine, followed by sequential treatment of the product with carbon disulphide, then methyl iodide, chloroacetone and potassium carbonate eventually gives the thiazine (176) (D.I.C. Scopes, A.F. Kluge and J.A. Edwards, J.org.Chem., 1977, *42*, 376). 2,6-Disubstituted- 4H-1,3-thiazin-4-ones may be synthesised by successive treatment of β-ketoamides (177) with perchloric acid and hydrogen sulphide (Y. Yamamoto, S. Ohnishi and Y. Azuma, Chem.pharm.Bull., 1983, *31*, 1929), and 2-mercapto analogues (178) of these structures are available through the reactions of 3-chloro-3-arylpropenoyl isothiocyanates and aliphatic thiols (J. Imrich and P. Kristian, Coll.Czech.chem.Comm., 1982, 47, 3268).

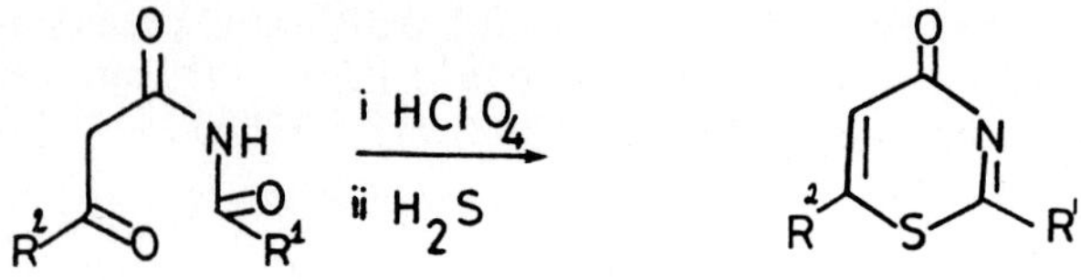

177

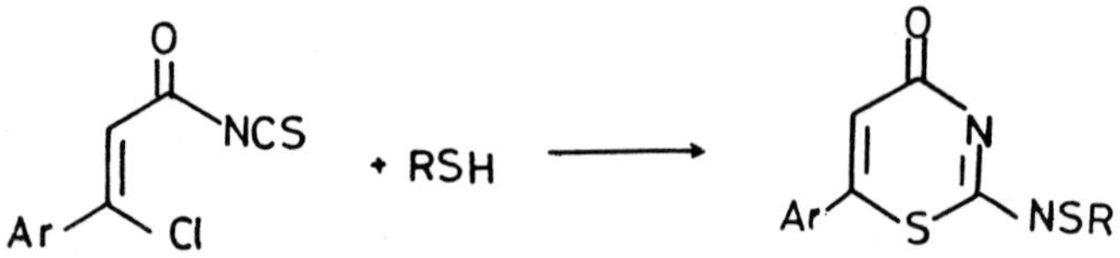

178

(ii) 1,3-Thiazinium salts

β-Chlorovinyl ketones and the thioamide (179), in the presence of perchloric acid, give rise to intermediate salts (180), which then cyclise to yield 1,3-thiazinium salts (W. Schroth et al., Tetrahedron, 1982, 38, 937).

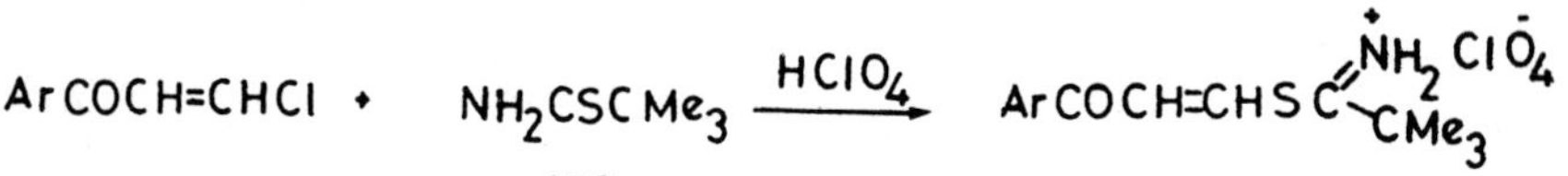

179

180

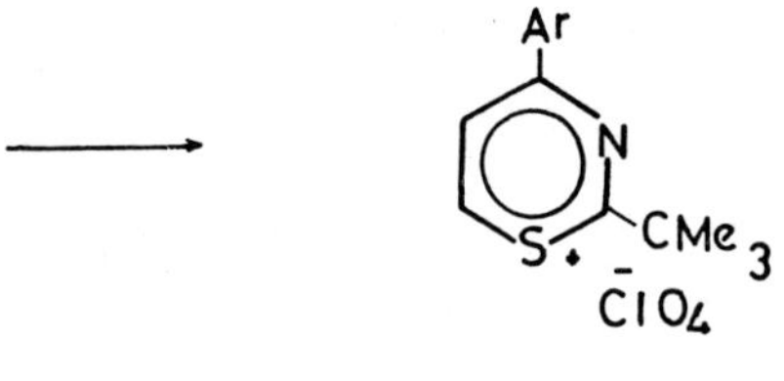

Normally 1,3-thiazinium salts react with nucleophiles at position 6 and on cathodic reduction they also undergo reductive dimerisation at this site to afford 6,6' bi-thiazoyls (181) (H.H. Ruttinger et al., J.prakt.Chem., 1981, 33).

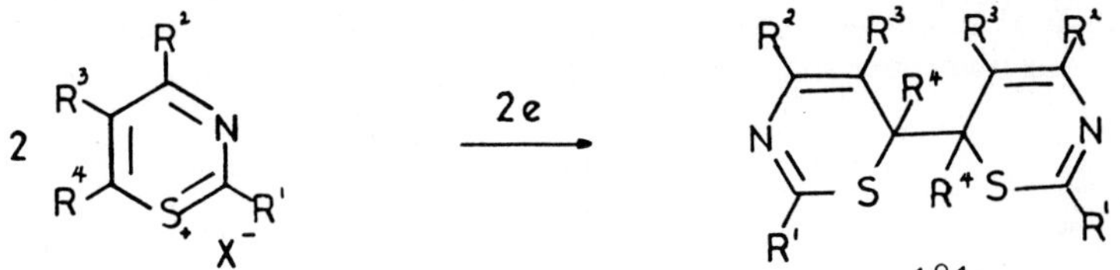

181

(iii) Dihydro-1,3-thiazines

These compounds are known in the form of oxo- and thioxo-derivatives, for example, ring expansion of the 2-substituted isothiazolone (182) with dimethyl diazomalonate in the presence of rhodium (II) acetate gives the 3,4-dihydro-1,3-thiazin-4(2H)-one (183) (W.D. Crow, I. Gozney and R.A. Ormiston, Chem.Comm., 1983, 643).

4H-5,6-Dihydro-1,3-thiazines are commonplace and can be obtained, for example, by the ring closure of 3-isocyano-mercaptans (184) in the presence of copper (I) oxide (V. Schollköpf, R. Jentsch and K. Madawindta, Ann., 1979, 451).

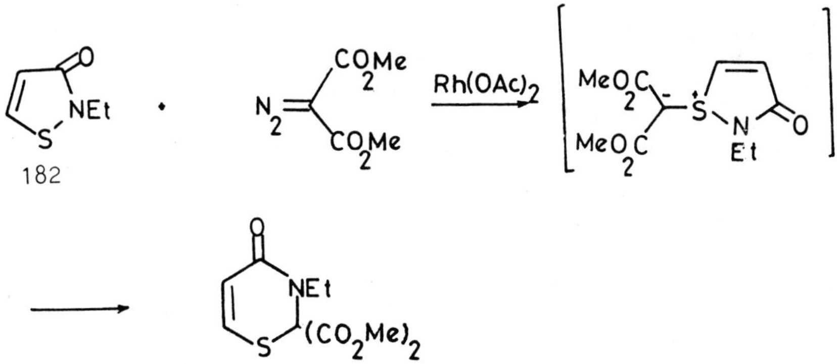

182

183

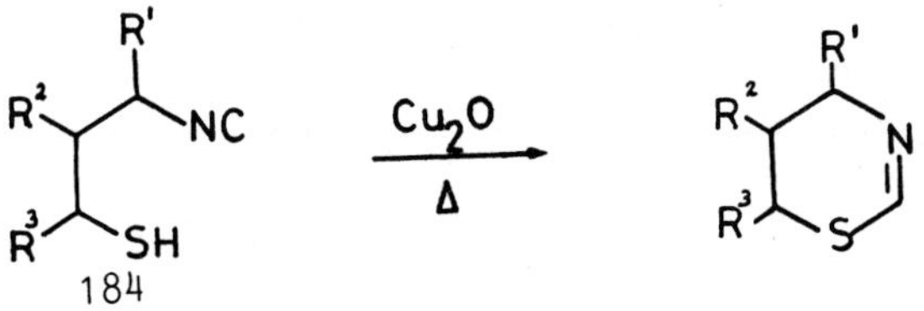

184

Similarly diazoketones (185), in the presence of acid, eliminate nitrogen and afford the 5-oxo-derivatives (186) (V.G. Kartsev et al., Khim.Geterot.Soedin., 1980, 1327).

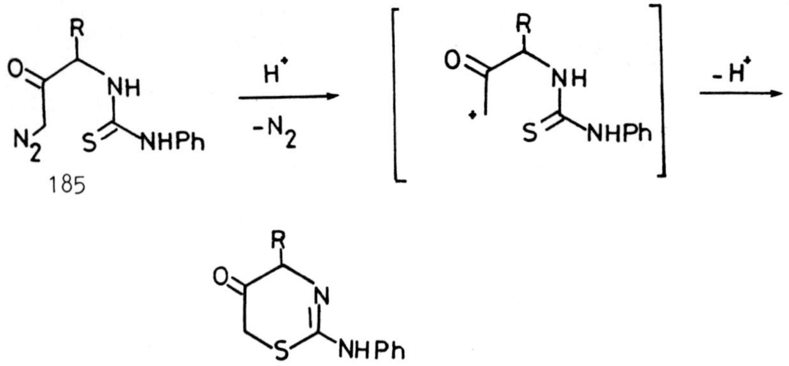

185

186

Aminoacrylonitriles cyclo-condense with carbon disulphide in dimethylformamide solution to form thiazinedithiones (187) (T. Yamamoto, S. Hirasawa and M. Muraoka, Bull.chem.Soc. (Japan), 1985, *58*, 771.
2-Substituted dihydro-1,3-oxazin-4,6-diones (188) are lithiated at position 5, and then react at this site when the enolates so formed are treated with alkyl halides (V.G. Beilin, E.N. Kirillova and L.B. Dashkevich, (Patent), C.A., 1978, *89*, 434413).

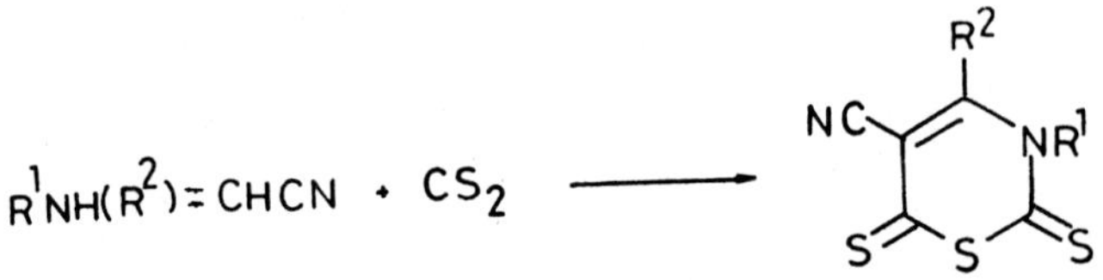

187

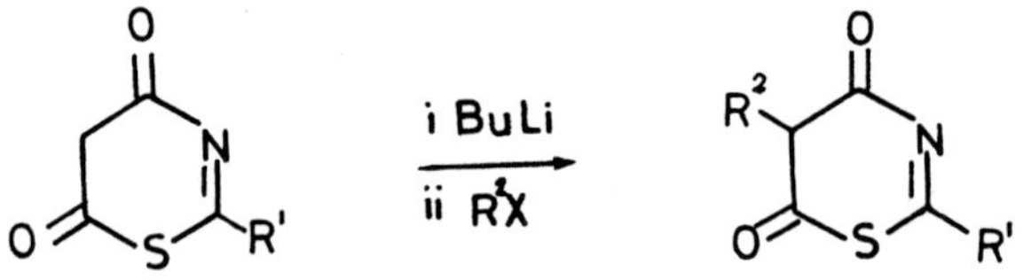

188

3,6-Dihydro-1,3-thiazine-2-thiones (190) arise when amines are treated with carbon disulphide sodium hydroxide and bromomethylacetylene, and the dithiocarbamates (189) thus obtained are cyclised with hydrochloric acid. However, aminomethylacetylenes react with carbon disulphide and triethylamine to give the isomeric 3,4-dihydro compounds (191) (W. Hanefeld, Arch.Pharm., 1984, *317*, 297).

$$RNHCS_2CH_2C{\equiv}CH \xrightarrow[\text{H}_2\text{O}]{\text{HCl}}$$

189

190

$$RNHCH_2C{\equiv}CH + CS_2 \xrightarrow{\text{Et}_3\text{N}}$$

191

Ethoxycarbonylacetylene and *N*-phenylthiourea cyclocondense to yield 2-phenylimino-2,3-dihydro-1,3-thiazin-4-one (192) (E.L. Khanina, D. Muceniece and G. Duburs, Khim.Geterot. Soedin, 1984, 997).

$$HC{\equiv}C\,COEt_2 + PhNHCSNH_2 \xrightarrow[\text{EtOH}]{\Delta}$$

192

2,6-Disubstituted-5,6-dihydro-4H-1,3-thiazin-4-ones are easily synthesised from *N*-substituted *N*'-(3-substituted-propenoyl) thioureas by treatment with boron trifluoride etherate, followed by basification with sodium bicarbonate (M. Dzurilla, P. Kutschy and P. Kristian, Synthesis, 1985, 933).

(iv) Tetrahydro-1,3-thiazines

Tetrahydro-1,3-thiazin-2-thiones can be made by reacting sodium dithiocarbamates with 1,3-dibromopropanes, or from thioureas and 1,3-dibromopropanes in the presence of sodium hydroxide and then treatment of the product imines with carbon disulphide (W. Hanefeld, Arch.Pharm., 1977, *310*, 409, Hanefeld and E. Bercin ibid., 1981, *314*, 413).

N-acyl derivatives of the thiones can be used as trans-acylating agents (Hanefeld and Bercin, ibid., 1984, *317*, 74), and the thiones themselves react with thionyl chloride to give 2,2-dichloro derivatives, or with amines, or ammonia to yield 2-imino compounds (idem. ibid., 1985, *318*, 60).

2-Thioxotetrahydro-1,3-thiazine-4-ones, unsubstituted at the nitrogen atom, are acylated at this position by acid chlorides. The *N*-acyl derivatives are thermally unstable (Hanefeld, Arch.Pharm., 1980, *313*, 833), and the parent structures are ring opened by amines to afford thioureas (E.V. Vladzimirskaya et al., C.A., 1978, *89*, 43283z). On the other hand, 2-imino derivatives (193) are obtained by the cycloaddition of acryloyl chloride to thioureas (S. Akhmedova, N.K. Rozhkova and R.F. Ambartsumova, Uzb.khim.Zhur., 1977, 59; Hanefeld, Arch.Pharm., 1977, *310*, 273).

193

In the case of tetrahydro-1,3-thiazine-2,6-diones oxidation with 3-chloroperbenzoic acid gives S-oxides and S,S-dioxides (Hanefeld, Ann., 1984, 1627).
The reactions of 1,3-thiazinedithiones have been well studied see, for example, M. Muraoka, A. Yamada and T. Yamamoto (J.heterocyclic Chem., 1984, *21*, 953).

(c) 1,4-Thiazines and reduced forms

(i) 1,4-Thiazines and dihydro-1,4-thiazines

2H-1,4-Thiazines are relatively uncommon although the parent heterocycle is well known; one method of preparation employs the sulphoxides formed from the oxidation of corresponding dihydrothiazines with 3-chloroperbenzoic acid. These compounds spontaneously dehydrate on standing to 2H-1,4-thiazines (194) (D.A. Berges and J.J. Taggart, J.org.Chem., 1985, *50*, 413; cf. A.G.W. Baxter and R.J. Stoodley, J.chem.Soc.Perkin I, 1976, 2540).

When treated with triethylamine alkyl 2H-1,4-thiazine 2,6-dicarboxylates (195) eliminate sulphur to give pyrroles, the yields are typically 58-77% (L.F. Lee and R.K. Howe, J.org.Chem., 1984, *49*, 4780).

4H-2,3-Dihydro-1,4-thiazines are available through many reactions, for example (i) by the addition of allenes (196) and 2-aminoethyl mercaptan in the presence of base (Z.T. Fomum et al., Heterocycles, 1982, *19*, 465); (ii) from the interaction of 2-chloroacetoacetic acid derivatives and the ethyl ester of cysteine (T. Chiba et al., ibid., 1984, *22*, 387); (iii) by the reaction of ketones with ethylenimine and sulphur (F. Asinger, J. Stalschus and A. Saus, Monatsch., 1979, *110*, 425). [N.B. In the last preparation the products (197) are contaminated with thiazolidines (198).]

A new route to 4H-2,3-dihydro-1,4-thiazines requires the reaction of β-bromoethylamine with alkyl α-alkylthio-β-keto-esters (199), followed by thermally promoted S-dealkylation of the derived salts (200). The whole process can be carried out as one step if the reactants are heated together in acetonitrile containing potassium carbonate (M. Hatanaka et al., Synthesis, 1985, 688).

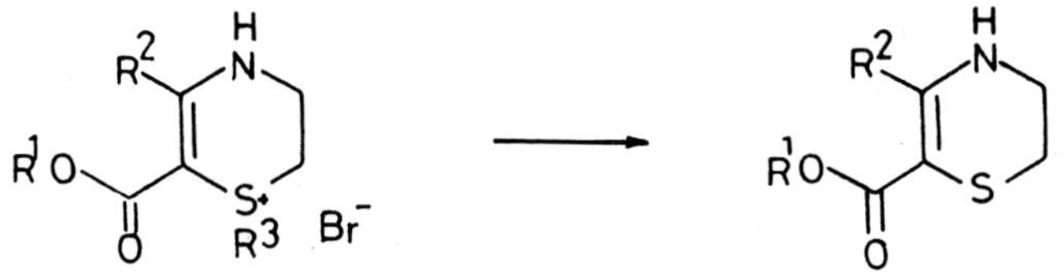

200

S-Oxides (202) are synthesised by the treatment of sulphoxides (201,n=1) or sulphones (201,n=2) with methyl-amine (W. Verboom, R.S. Sukhai and J. Meijer, Synthesis, 1979, 47).

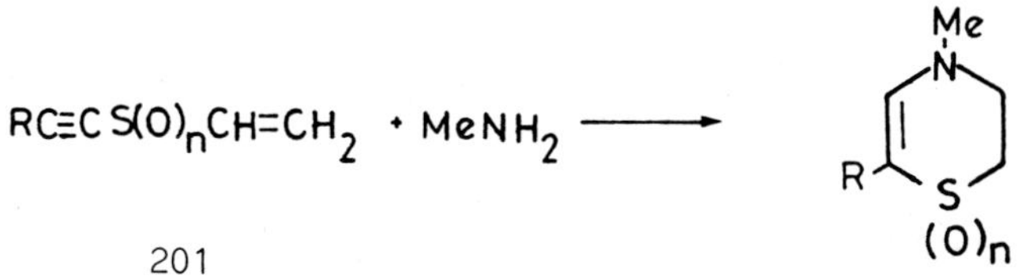

201

202

4-Alkyl-2H-1,4-thiazin-3-ones (203) are prepared by reacting 2-iodomethylthiazolium iodides with potassium hydroxide (M. Hojo et al., Synthesis, 1979, 272).
In general 2H-1,4-thiazin-3-ones are cleaved to N-acetyl-3-mercaptoethenamines by treatment with sodium or lithium in liquid ammonia (A.J.G. Baxter, R.J. Ponsford and R. Southgate, Chem.Comm., 1980, 429). With 3-chloroper-benzoic acid oxidation of the sulphur atom is not always observed, instead the esters (204) may form (Hojo et al., Synthesis, 1982, 312).

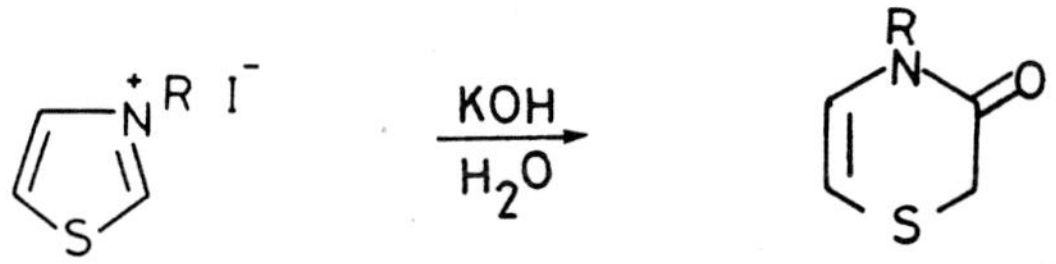

203

MCPBA

204

In view of this *S*-oxides are best synthesised by the ring-closure of substrates in which there are already sulphoxide or sulphone functions present (see M. Bobek, J.heterocyclic Chem., 1982, *19*, 131).

(ii) Tetrahydro-1,4-thiazines and thiazinones

Oxirane carbonitrile (205) and cysteine methyl ester combine together to give the tetrahydro-1,4-thiazine (206) (J. Kopecky et al., Tetrahedron Letters, 1984, *25*, 4295). 3-Carboxylic ester derivatives can also be prepared in a conventional manner by reacting cysteine esters with α-halo-ketones, followed by reduction of the products with sodium borohydride (N. Yoneda and K. Sakai (Patent), C.A., 1980, *93*, 204671z). *S,S*-Dioxides (208) are available either by oxidation of thiomorpholines (F. Asinger, A. Saus and M. von Wachtendonk, Monatsch., 1980, *111*, 385), or from dialkylethan-1,2-disulphonyl acetates (207) and aromatic aldehydes in the presence of ammonium acetate (R. Jeyaraman, C.D. Jayaraj and M. Chockalingam, Ind.J.Chem., 1981, *20B*, 333).

HCl

205   **206**

$$CH_2\,S(O)_2CH_2CO_2Et$$
$$CH_2\,S(O)_2CH_2CO_2Et$$
$$\xrightarrow[\text{NH}_4\text{OAc}]{\text{ArCHO}}$$

207   208

Tetrahydro-1,4-thiazine in the form of the *N,N*-dimethyl salt (210) is obtained through the cyclisation of the sulphoxide (209) with mineral acids (B.A. Trofimov et al., (Patent), C.A., 1980, *92*, 6545c).

Syntheses of tetrahydro-1,4-thiazin-3-ones follow conventional routes to the parent structures, except that α-chloroketones are replaced by α-chloroacids. As an illustration, reactions of chloroacetic acid and amino thiols (211) yield 4,6-disubstituted compounds (212) (M.A. Allakhverdiev et al., Khim.Geterot.Soedin., 1984, 327).

$$\xrightarrow{\text{HX}}$$

209   210

$$\xrightarrow{\text{ClCH}_2\text{CO}_2\text{H}}$$

211   212

5. Benzothiazines

(a) 1,2-, 2,1- and 2,3-Benzothiazines and related structures

2H-1,2-Benzothiazin-4(3H)-ones are often represented as the tautomeric enols (213). J.G. Lombardino (Org. Prep.

Proced.Int., 1980, *12*, 269) states that alkylation occurs primarily at the nitrogen atom in such compounds, but this preference could simply depend on the "hardness or softness" of the reagent.   In the case of the 3-acyl derivatives (R=COPh) nucleophiles, e.g. Grignard reagents, amines etc., react at the exocyclic carbonyl group (N.M. Abed, Ind.J.Chem., 1976, *14B*, 428).

213

These compounds can be synthesised by the cyclisation of *N*-(2-alkoxycarbonylbenzenesulphonyl)glycinates   (214) (Lombardino (Patent) C.A., 1977, *86*, 5478x), or by base mediated ring expansion of 3-bromoalkyl-1,2-benzisothi-azole-1,1-dioxides (215) (R.A. Abramovitch et al., Chem.Comm., 1976, 771).

An alternative pathway involves the ring-opening/recyclisa-tion of the saccharin derivative (216) with sodium hydride in either dimethylformamide or dimethylsulphoxide (P.D. Weeks et al., J.org.Chem., 1983, *48*, 3601).

214   215

216

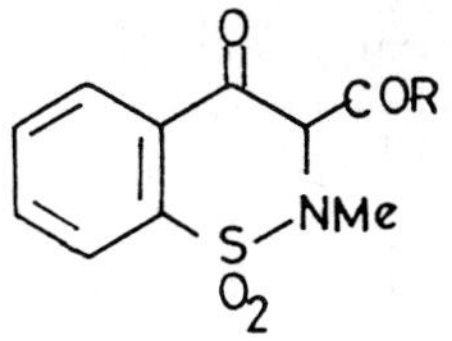

217

Piroxicam (217,R=2-aminopyridyl) is an effective anti-inflammatory drug.

When treated with trifluoroacetic acid, or with boron trifluoride in acetic acid, the oxime of 2-methyl-2H-1,2-benzothiazin-4-(3H)-one rearranges to the benzothiazin-3(4H)-one (218) by a mechanism which seems to resemble that of the Sommler Wolff transformation of α-tetralone oxime to α-naphthylamine (H. Zinnes, R.A. Comes and J. Shavel Jr., J.heterocyclic Chem., 1977, 14, 1063).

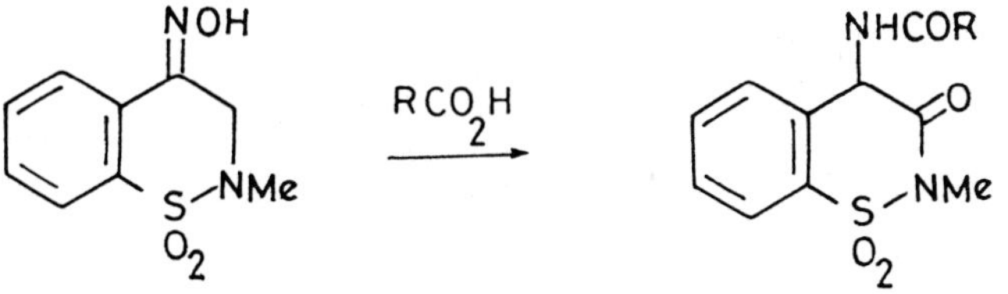

218

6H-Dibenzo[c,e][1,2]thiazine (220) is synthesised by cyclodehydrohalogenation of the sulphonamide (219) with palladium (II) acetate (D.E. Ames and A Opalko, Tetrahedron, 1984, 40, 1919), whereas the tetracyclic structure (222) is obtained by heating the salt (221) in carbon tetrachloride solution (E.K. Adesogan and B.I. Alo, Chem.Comm., 1979, 673).

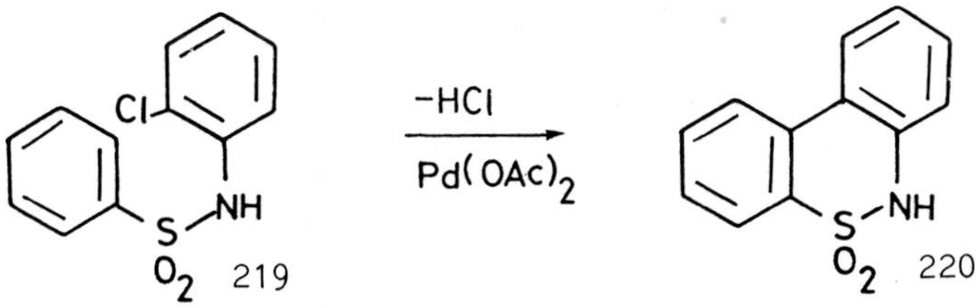

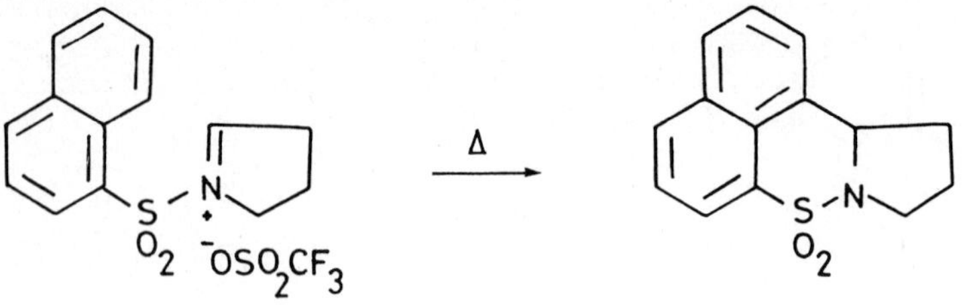

221   222

The pyrido [2,3 c][1,2]thiazine (224) is formed in two steps
by reacting 2-chloronicotinonitrile with N-methylmethyl-
sulphonamide and treatment of the intermediate product (223)
with base (G.M. Coppola and G.E. Hardtmann, J.heterocyclic
Chem., 1979, 16, 1361).

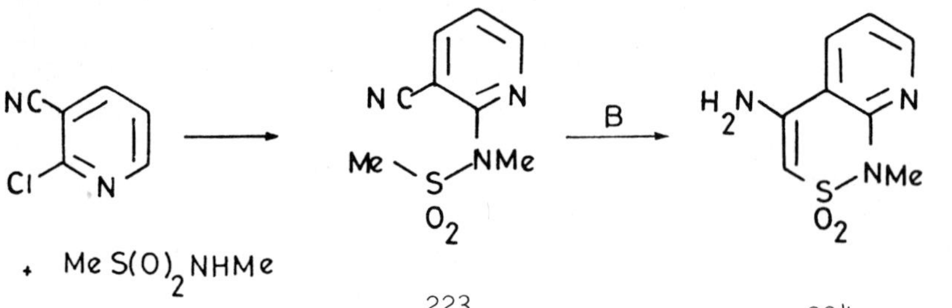

223

224

Vacuum flash pyrolysis of 2,6-dichloro-β-phenethylsulphonyl-
azide (225) affords 5,8-dichloro-3,4-dihydro-2,1-benzo-
thiazine-2,2-oxide (226) resulting from a 1,2-chlorine
shift, possible as shown below (R.A. Abramovitch et al.,
J.org.Chem., 1985, 50, 2066). The 2,6-dimethyl analogue
behaves similarly, but now gives 5,8-dimethyl-3,4-dihydro-
2,1-benzothiazine-2,2-oxide through a 1,2-methyl migration.

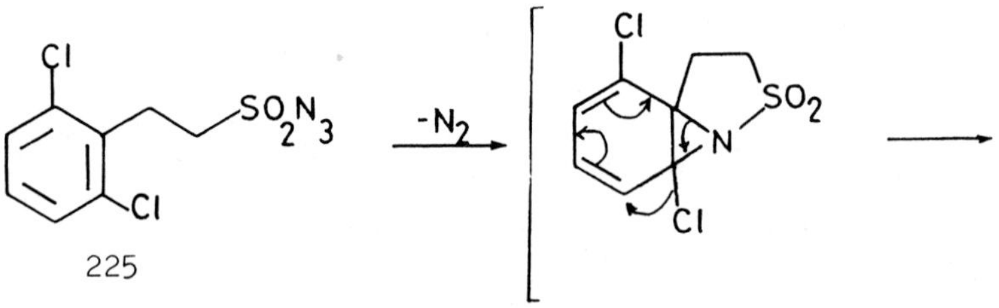

225

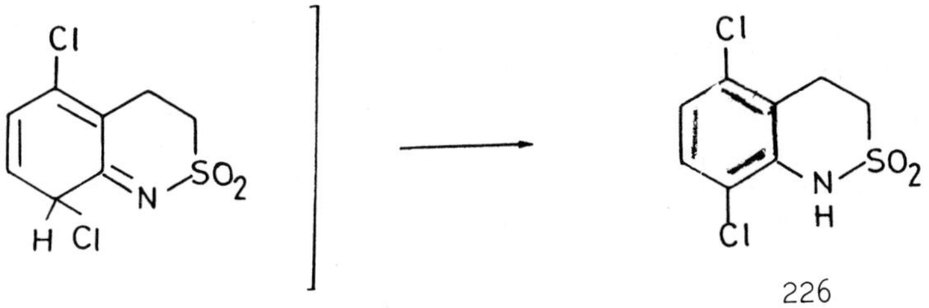

226

2,1-Benzothiazinium ylides (227) are produced by the cyclisation of β-mercaptostryenes with *N*-chlorosuccimide (NCS) and potassium hydroxide (two steps). With hydrogen halides these compounds protonate at nitrogen affording 1,2-dihydro-2,1-benzothiazines (228) (M. Hori et al., Tetrahedron Letters, 1979, 3969).

A ring expansion reaction occurs when the thiophthalylium salt (229) is heated with hydroxylamine hydrochloride in acetic acid, followed by treatment with acetic anhydride and perchloric acid. The product is the 1H-2,3-benzothiazinium-perchlorate (230) (D.A. Oparin, T.G. Melent'eva and L.A. Pavlova, Zh.org.Khim., 1983, *19*, 1986).

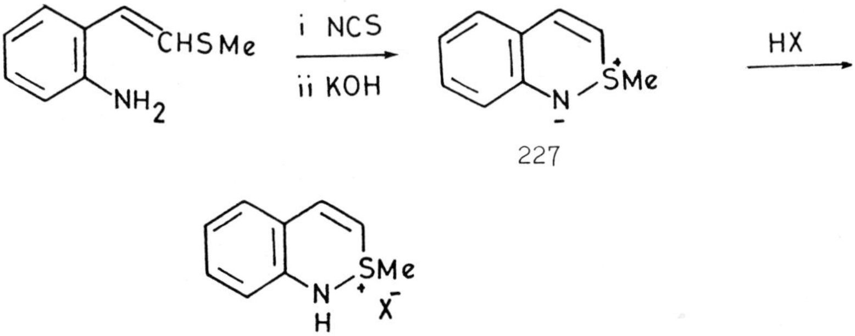

227

228

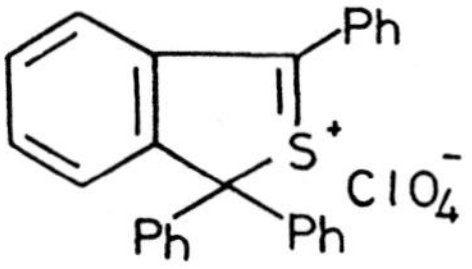

229

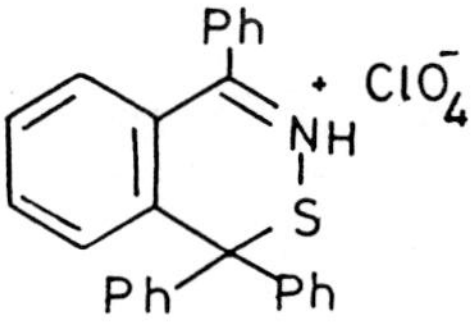

230

(b) 1,3-Benzothiazines and reduced forms

(i) 2H- and 4H-1,3-Benzothiazines

These structures may be synthesised by a modified Ritter reaction, thus aryl chloromethyl sulphides react with nitriles in the presence of antinomy (V) chloride to give 63-80% yields of the heterocycles (231) (D.K. Thakur and D. Yashwant, Synthesis, 1983, 223).

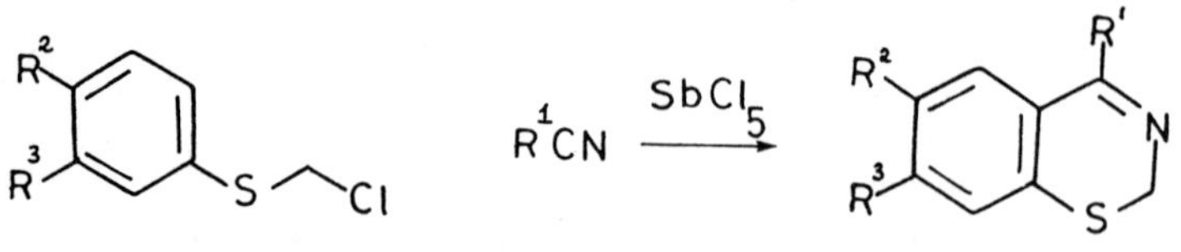

231

In the phosphorus oxychloride cyclisations of the amides (232) yields are poor, and mixed 2H- and 4H-1,3-benzo-thiazines are formed (J.Szabo et al., Pharmazie, 1984, *39*, 426). Better yields are obtained through the copper(I) catalysted intramolecular cyclisation of iodothioamides (233,R=H or OH) (W.R. Bowman, H. Heany and P.H.G. Smith, Tetrahedron Letters, 1984, *25*, 821).

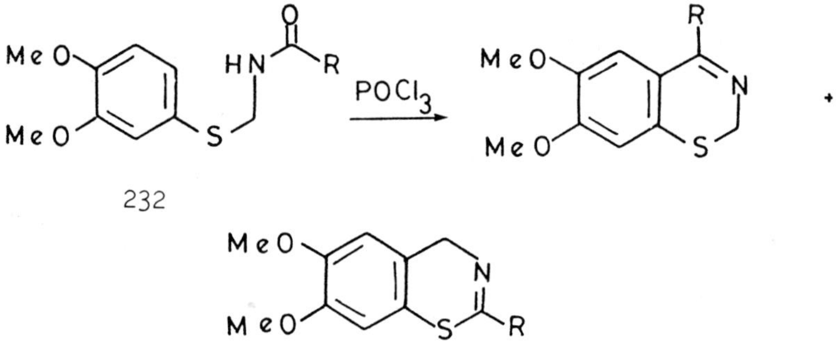

232

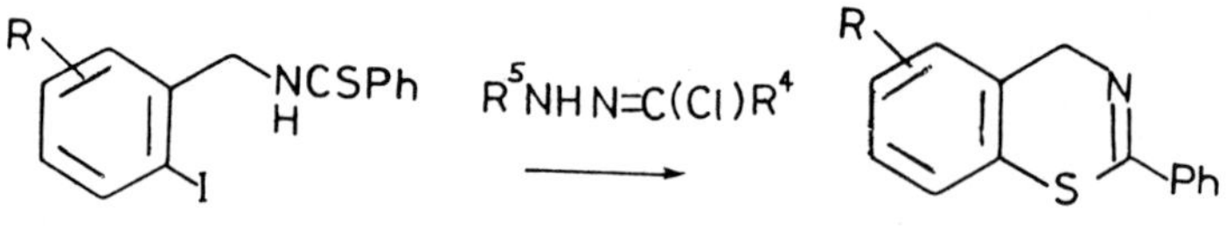

233

2 H-1,3-Benzothiazines undergo 1,3-dipolar addition reactions, and with nitrilimines for example, triazolobenzo-thiazines (234) are formed (L. Fodor et al., Heterocycles, 1984, 22, 537).

6,7-Dimethoxy-2H-1,3-benzothiazines (235,R^1=H or Ph) react with substituted acetyl chlorides to form angularly condensed β-lactams (236). However, when R^1=methyl this substrate furnishes enamines (237) (P. Sohar et al., Tetrahedron, 1984, 40, 4387; see also p.4089).

Reduction of 4H-1,3-benzothiazines with zinc and acetic acid affords the corresponding 2H-3,4-dihydro derivatives (T.V. Sokolova et al., Khim. Farm.Zhur., 1976, 10, 42).

234

235

236

237

(c) 3,1-Benzothiazines and related forms

4H-3,1-Benzothiazines (239) are prepared by the cyclo-additions of thioketones and ketenimines (238) (A. Dondoni, A. Battaglia and P. Giorgianni, J.org.Chem., 1980, 45, 3766), whereas 2-alkylidene-benzothiazin-4-ones (241) are produced by reacting 2-amino-benzoic acids or esters with dithietanes (240) (K. Peseke, Synthesis, 1976, 386).

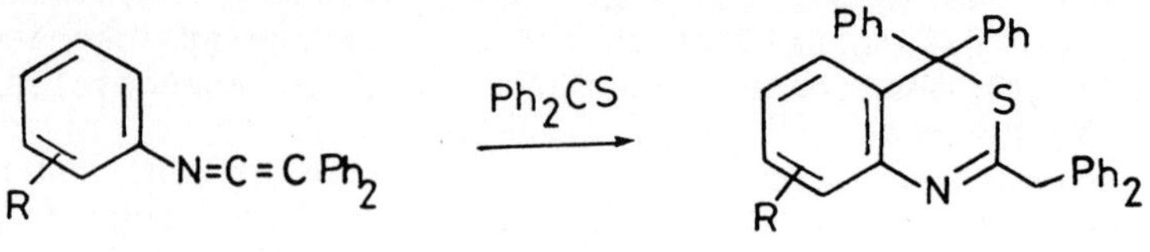

238

Ph$_2$CS

239

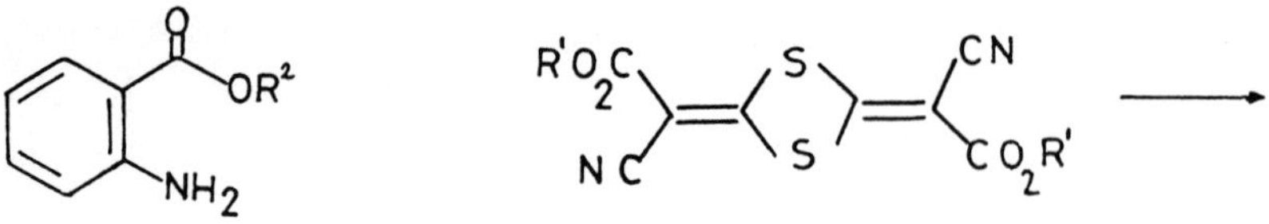

240

241

Naphtho-[1,2 d][3,1]thiazines result from the reaction of α-tetralones with thiourea in contact with hydrochloric acid in ethanol. The initial products are iminium salts (242), which on treatment with ammonia tautomerise to amines (243) (T. Lorand et al., Pharmazie, 1984, *39*, 536).

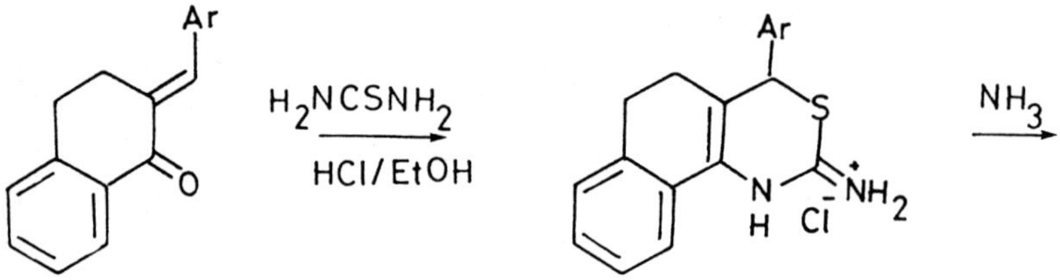

H$_2$NCSNH$_2$

HCl/EtOH

NH$_3$

242

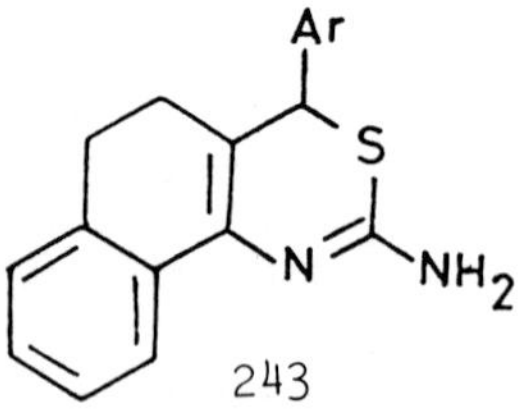

243

A simple synthesis of the tricyclic dihydrobenzothiazines (246) is furnished by reacting the pyrrolidinyl ketones (244) with Lawesson's reagent (245) (W. Verboom et al., Tetrahedron Letters, 1984, 25, 4309).

The chemistry of 3,1-benzothiazine-2,4-dithione has been extensively studied by S. Leistner and G. Wagner (Pharmazie, 1975, 30, 542; 1980, 35, 124; 293; Z.Chem., 1984, 24, 328). It has been observed that amines react with these compounds to yield quinazolinthiones, whereas cyanogen bromide affords 2-thiocyanato-3,1-benzothiazin-4-thiones .

(d) 1,4-Benzothiazines

(i) 4H-1,4-Benzothiazines
4H-1,4-Benzothiazines can be synthesised in a number of ways. Conventionally (a) by the cycloaddition of 2-amino-thiophenols and alkynes (G. Liso et al., J.heterocyclic Chem., 1980, *17*, 377; 793) and (b) by reacting the former with α,β-dicarbonyl compounds (S.K. Jain and R.L. Mital, J.Inst.Chem.India, 1977, *49*, 282; R. Soni and M.L. Jain, Tetrahedron Letters, 1980, *21*, 3795; R.R. Gupta et al., 1980, *14*, 1145; Bull.Chem.Soc.(Japan), 1984, *57*, 2343; Heterocycles, 1984, *22*, 1143). Less conventional methods (although quite efficient) include the rearrangement of 2-alkyldihydrobenzothiazoles (247) with sulphuryl chloride (F. Chioccara et al., Chim.Ind.(Milan), 1976, *58*, 546; Synthesis, 1977, 876) or, where the 2-substituent bears an "active methylene" group, potassium *t*-butoxide (G. Liso et al., ibid., 1983, 755). Dihydrobenzothiazole-*S*-oxides (247,R=0) may be induced to undergo ring expansion with 4-toluene-sulphonic acid, or acetic anhydride to 1,4-benzo-thiazine-*S*-oxides (Chioccara, Synthesis, 1978, 744; M. Hori et al., Tetrahedron Letters, 1981, *22*, 1701). In some instances the products from ring expansion reactions of this type are dihydro-1,4-benzothiazines bearing an alkylidene unit at C-3 (see Hori et al., C.A., 1980, *93*, 71673c).

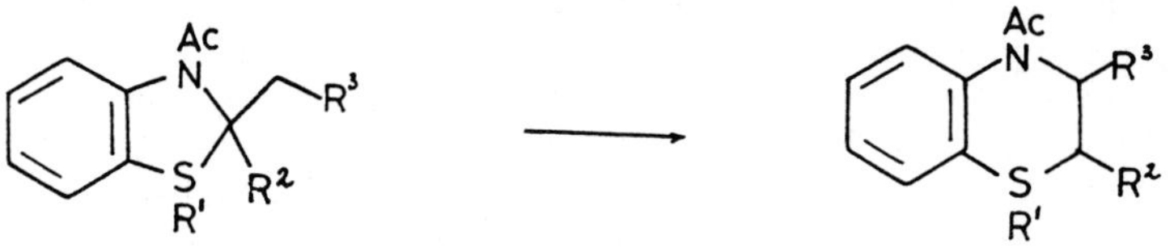

247

Benzyl-2-aminosulphides (248) can be cyclised by treatment
with lithium diisopropylamide to give benzothiazines
(F. Badudri et al., J.chem.Soc.Perkin I, 1984, 1899) and
zwitterionic species (250) are formed by the ring-closure of
the sulphoxides(249) with trifluoroacetic anhydride
(T.L. Gilchrist and G.M. Iskander, idem., 1984, 1899).

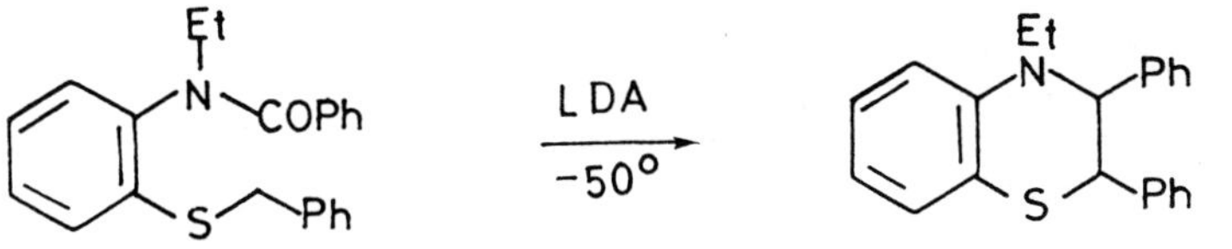

248

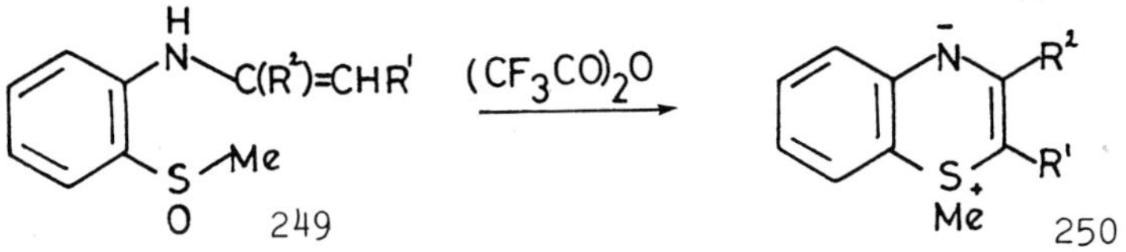

249   250

2H-1,4-Benzothiazine itself is highly unstable and undergoes
reversible aldolization to give mainly the cyclic trimer
(251) (Chioccara et al., Chem.Comm., 1977, 50).

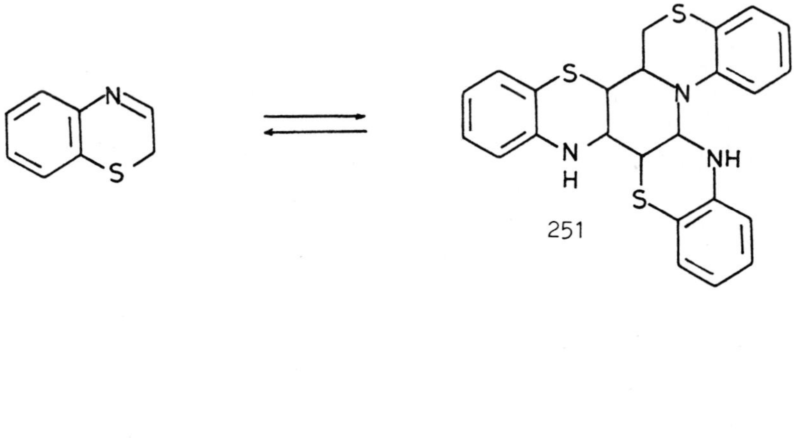

251

(ii) 2,3-Dihydro-4H-1,4-benzothiazines

If benzothiazolines (252) are reacted with bromocarbonyl compounds ring scission occurs to produce imines (253), which on base treatment re-cyclise to 2,3-dihydro-4H-1,4 benzo-thiazines (S.H. Mashraqui and R.M. Kellogg, Tetrahedron Letters, 1985, *26*, 1457). Sulphones (255) are synthesised by treating imines (254) with methanesulphonyl chloride and triethylamine (M. Rai et al., Chem.Ind., 1979, 26).

Another route involves the treatment of *N*-allylanilines (256) with sulphur (II) chloride (M. Mühlstadt, K. Hollmann and R. Wider, Tetrahedron Letters, 1983, *24*, 3203).

Unsubstituted   4-acyl-2,3-dihydro-4H-1,4-benzothiazines undergo ring fission when treated with lithium di-isopropylamide (LDA) giving 2-(alkylthio)phenylenamides and other products (F. Badudri et al., J.org.Chem., 1983, *48*, 4082).

Butyllithium on the other hand effects partial *N*-deacetylation (F. Ciminale, L. Dinunno and S. Florio, Tetrahedron Letters, 1980, *21*, 3001).

N-Alkyl-2,3-dihydro-4H-1,4-benzothiazin-3-ones   may   be metallated with LDA at position 3 and then condensed with aldehydes to afford aldol products which dehydrate to 3-alkylidene derivatives (Badudri et al., Tetrahedron, 1982, *38*, 3059; 1985, *41*, 569).   Sodium hydride reacts similarly and the anion so formed is also capable of reacting with esters to yield 3-acylbenzothiazinones (F. Eiden and F. Meinel, Arch.Pharm., 1979, *312*, 302).

6.   Phenothiazines

A common synthesis of phenothiazines utilizes the Turpin reaction of 2-nitro-2'-mercaptodiphenylamines (see C.C.C. 2nd Edn., Vol. IV^H Chap.41, p.517).   This continues to be a widely used approach despite the possibility of Smiles rearrangements leading to mixed products (R.C. Chaudhary, Ann.Soc.Sci.Bruxelles Ser.1, 1979, *93*, 253;   V.N. Knyazev, V.N. Drozd, T.Ya Mozhaeva, Zhur.org.Khim., 1979, *15*, 2561;

R.R. Gupta and R. Kumar, Heterocycles, 1984, *22*, 1169;
R. Schneider and A. Buge, Pharmazie, 1984, *39*, 22).
S,S¬Dioxides (258) are available through triethyl phosphite
reductions of nitrodiphenylsulphones (257), and again spiro -
intermediates are probably involved, with attendant
implications for isomer production (J.I.G. Cadogan et al.,
J.chem.Soc.Perkin I, 1976, 1749).

257

258

1- Alkylphenothiazines are readily produced by the treatment
of 2- chlorophenothiazine with alkyllithiums.   This general
procedure can also be used in forming 1- phenylphenothiazine
using phenyllithium as the reagent.   Here it is assumed
that an aryne intermediate is involved leading to addition
of the reagent across the 1,2 bond.   Finally acidification
releases the phenothiazine (A. Hallberg et al.,
J.chem.Soc.Perkin I, 1985, 969).
2,3-Dihydro-1H-phenothiazin-4 -(10H)- ones (260) are obtained
by cyclising mercaptoanilines (259) with diones (M.L. Jain
and R.D. Soni, Synthesis, 1983, 933).

259

260

1-Acetylphenothiazines are prepared in low yield by the photochemically induced rearrangement of 10-acetylpheno-thiazines (Y. Ueno, Pharmazie, 1983, *38*, 567); similar changes occur on pyrolysis, but now the products are 3-acetylphenothiazines (idem. ibid., 1984, *39*, 582).

Similarly, 10-sulphonylated phenothiazines rearrange to 3-sulphonyl derivatives when heated with sodium hydroxide in dimethylformamide (L.S. Kapishchenko and S.I. Burmistrov, Khim.Geterot.Soedin., 1976, 1365). 10-Acylphenothiazines undergo *S*-imidation with *O*-(mesitylsulphonyl)-hydroxylamine (P. Stoss and G. Satzinger, Ber., 1978, *111*, 1453).

Halogenated phenothiazin-3(3H)-ones are synthesised by treating zinc 2-aminophenylmercaptides with 1,4-benzo-quinones (S.K. Jain and R.L. Mital, Rev.Roum.Chim., 1980, *25*, 697; A.R. Goyal and Mital, Gazz., 1980, *110*, 205; R.R. Gupta et al., Ann.Soc.Sci.Bruxelles Ser.1, 1983, *97*, 101; M.H. Terdic, Rev.Roum.Chim., 1984, *29*, 489).

Zinc dust, or sodium dithionite, reduces phenothiazin-3-ones to 3-hydroxyphenothiazines (Jain and N.L. Agarwal, Z.Naturforsch. 1980 , *35B*, 381).

7.   Phenotellurazines

This new heterocyclic system is represented by the derivative (262) which can be obtained by the reaction of the bromoaniline (261) with butyllithium and tellurium iodide.   The mechanism of this unusual reaction *may* involve the addition of a methylbenzyne to an intermediate aryltellurium derivative.   The tricycle undergoes halogenation and methylation at the tellurium atom (I.D. Sadekov et al., Dokl.Akad.Nauk.S.S.S.R., 1982, *266*, 1164).

261   262

Guide to the Index

This index is constructed in a similar manner to the volume indexes of the first edition of the Chemistry of Carbon Compounds. However, to make the index easier to use, more descriptive entries have been made for the commonly occurring individual, and groups of chemicals.

The indexes cover primarily the chemical compounds mentioned in the text, and also include reactions and techniques, where named, and some sources of chemical compounds such as plant and animal species, oils, etc.

Chemical compounds have been indexed alphabetically under the names used by authors, editing being restricted to ensuring uniformity of entries under the same heading. In view of the alternative nomenclature that can often be used, a limited amount of cross-referencing has been done where it is considered to be helpful, but attention is particularly drawn to Convention 2 below.

For this and the succeeding volumes, the indexing conventions listed below have been adopted.

1. *Alphabetisation*

(a) The following prefixes have not been counted for alphabetising:

n-	*o-*	*as-*	*meso-*	D	C
sec-	*m-*	*sym-*	*cis-*	DL	*O-*
tert-	*p-*	*gem-*	*trans-*	L	*N-*
	vic-				*S-*
		lin-			*Bz-*
					Py-

Some prefixes and numbering have been omitted in the index, where they do not usefully contribute to the reference.

(b) The following prefixes have been alphabetised:

Allo	Epi	Neo
Anti	Hetero	Nor
Cyclo	Homo	Pseudo
	Iso	

(c) A letter by letter alphabetical sequence is followed for entries, firstly for the main entry, followed by the descriptive entry. The only exception to this sequence is the placing of plural entries in front of the corresponding individual entries to prevent these being overlooked by a strict alphabetical sequence which could lead to a considerable separation of plural from individual entries. Thus "butanes" will come before *n*-butane, "butenes" before 1-butene, and 2-butene, etc.

2. *Cross references*

In view of the many alternative trivial and systematic names for chemical compounds, the indexes should be searched under any alternative names which may be indicated in the main body of the text. Only a limited amount of cross-referencing has been carried out, where it is considered that it would be helpful to the user.

3. *Esters*

In the case of lower alcohols esters are indexed only under the acid, e.g. propionic methyl ester, not methyl propionate. Ethyl is normally omitted e.g. acetic ester.

4. *Derivatives*

Simple derivatives are not normally indexed if they follow in the same short section of the text.

5. *Collective and plural entries*

In place of "– derivatives" or "– compounds" the plural entry has normally been used. Plural entries have occasionally been used where compiunds of the same name but differing numbering appear in the same section of the text.

6. *Main entries*

The main entry of the more common individual compounds is indicated by heavy type. Multiple entries, such as headings and sub-headings over several pages are shown by "–", e.g., 67–74, 137–139, etc.

INDEX

270